Basic Structures
for Engineers
and Architects

Basic Structures
for Engineers and Architects

Philip Garrison
BSc, MBA, CEng, MICE, MIStructE, MIHT, ILTM
School of the Built Environment
Leeds Metropolitan University

Blackwell
Publishing

© 2005 Philip Garrison

Blackwell Publishing Ltd
Editorial offices:
Blackwell Publishing Ltd, 9600 Garsington Road, Oxford OX4 2DQ, UK
 Tel: +44 (0)1865 776868
Blackwell Publishing Inc., 350 Main Street, Malden, MA 02148-5020, USA
 Tel: +1 781 388 8250
Blackwell Publishing Asia Pty Ltd, 550 Swanston Street, Carlton, Victoria 3053, Australia
 Tel: +61 (0)3 8359 1011

The right of the Author to be identified as the Author of this Work has been asserted in accordance with the Copyright, Designs and Patents Act 1988.

All rights reserved. No part of this publication may be reproduced, stored in a retrieval system, or transmitted, in any form or by any means, electronic, mechanical, photocopying, recording or otherwise, except as permitted by the UK Copyright, Designs and Patents Act 1988, without the prior permission of the publisher.

First published 2005 by Blackwell Publishing Ltd

Library of Congress Cataloging-in-Publication Data

Garrison, Philip, CEng.
 Basic structures for engineers and architects / Philip Garrison.
     p. cm.
 Includes bibliographical references and index.
 ISBN 1-4051-2053-3 (pbk. : alk. paper)
 1. Structural engineering. I. Title.

 TA637.G37 2005
 624.1--dc22
                                                                                    2004023324

ISBN-10 1-4051-2053-3
ISBN-13 978-14051-2053-1

A catalogue record for this title is available from the British Library

Set in 9.5/12 pt Palatino
by Sparks, Oxford – www.sparks.co.uk
Printed and bound in India
by Replika Press Pvt. Ltd, Kundli

The publisher's policy is to use permanent paper from mills that operate a sustainable forestry policy, and which has been manufactured from pulp processed using acid-free and elementary chlorine-free practices. Furthermore, the publisher ensures that the text paper and cover board used have met acceptable environmental accreditation standards.

For further information on Blackwell Publishing, visit our website:
www.thatconstructionsite.com

To Jenny B

Contents

	Introduction	ix
	Acknowledgements	xv
1	What is structural engineering?	1
2	Learn the language: a simple explanation of terms used by structural engineers	10
3	How do structures (and parts of structures) behave?	13
4	Force, mass and weight	29
5	Loading – dead or alive	36
6	Equilibrium – a balanced approach	43
7	More about forces: resultants and components	49
8	Moments	60
9	Reactions	70
10	Different types of support – and what's a pin?	77
11	A few words about stability	86
12	Introduction to the analysis of pin-jointed frames	100
13	Method of resolution at joints	105

14	Method of sections	126
15	Graphical method	134
16	Shear force and bending moments	145
17	This thing called stress	175
18	Direct (and shear) stress	181
19	Bending stress	191
20	Combined bending and axial stress	214
21	Structural materials: concrete, steel, timber and masonry	229
22	More on materials	242
23	How far can I span?	249
24	Calculating those loads	258
	Appendices	269
	Index	277

Introduction

When I was 16 I had a Saturday job as a shelf-stacker at a local supermarket. One day, during a tea break, a co-worker asked me what I did the rest of the week. I explained that I had just done O Levels and was going on to do A Levels. I told him how many and in which subjects. He then asked me about my career aspirations (not his exact words). I explained that I wanted to become an engineer. His aghast response was: 'What! With all those qualifications?'

Engineers suffer from a lack of public perception of what their profession entails – many people think we spend our days in the suburbs, mending washing machines and televisions. Architects are more fortunate in this respect – the public have a better grasp of their profession: 'They design buildings, don't they?'

Public perceptions aside, careers in both civil engineering and architecture can be extremely rewarding. There are few other careers where individuals can be truly creative, often on a massive scale. The civil engineering profession offers a variety of working environments and a large number of specialisms within civil engineering. Civil engineers have opportunities to work all over the world, on projects large and small, and could come into contact with a wide variety of people, from the lowest worker on a construction site to government officials and heads of state.

At the start of the 21st century there is a huge demand for civil engineers and many young people (and some not so young!) are realising that this is a profession well worth entering.

Traditionally, students embarking on university courses in civil engineering would have A Levels in subjects such as mathematics, physics and chemistry. However, for a variety of reasons, many of today's potential students have A Levels (or similar) in non-numerate and non-scientific subjects. Moreover, a sizeable number of 'mature' people are entering the profession following a first career in something completely different. As a university admissions tutor, I speak to such people every day. It is possible, depending on the specialism eventually chosen, to enjoy a successful career in civil engineering without an in-depth mathematical knowledge. However, it is extremely difficult to obtain a degree or HND in civil engineering without some mathematical proficiency.

Turning to architects – these are creative people! Every building they design has a structure, without which the building would not stand up. Architects, like civil engineers, have to understand the mechanisms which lead to successful structures.

This book is about Structures. Structures is a subject studied as part of all civil engineering degree, HND and OND courses, as well as architecture degree courses, and also on some degree courses in related subjects (e.g. quantity surveying, building surveying, construction management and architecture).

The purpose of this book

I have taught Structures to undergraduate civil engineers and architects for the past 12 years. During that time I have noticed that many students find the basic concept of structures difficult to grasp and apply.

This book aims to do the following:

- to explain structural concepts clearly, using analogies and examples to illustrate the points;
- to express the mathematical aspects of the subject in a straightforward manner that can be understood by mathematically weak students and placed in context with the concepts involved;
- to maintain reader interest by incorporating into the text real-life examples and case histories to underline the relevance of the material that the student is learning.

This book presumes no previous knowledge of structures on the part of the reader. It does, however, presume that the reader has a good general education and a mathematical ability up to at least GCSE standard.

The intended readership

This book is aimed at:

- National Certificate (ONC), National Diploma (OND), Higher National Certificate (HNC), Higher National Diploma (HND) or first-year degree (BSc, BEng or MEng) students on a civil engineering (or similar) course, who will study a module called Structures, Structural Mechanics, Mechanics or Structural Analysis;
- students on a BA degree course in Architecture.

The following will also find this book useful:

- students on courses in subjects related to civil engineering and architecture – e.g. Quantity Surveying, Building Surveying, Construction

Management or Architectural Technology – who have to do a Structures module as part of their studies;
- those studying Technology at GCE A Level, GNVQ or AVCE;
- people working in the construction industry in any capacity.

The following will find the book a useful revision tool:

- a second (or subsequent) year student on a Civil Engineering or Architecture degree;
- a professional in the civil engineering or building industry, and practising architects.

A word about computers

Computer packages are available for every specialism and structural engineering is no exception. Certainly, some of the problems in this book could be solved more quickly using computer software. However, I do not mention specific computer packages in this book and where I mention computers at all, it is in general terms. There are two reasons for this.

(1) The purpose of this book is to acquaint the reader with the basic principles of structures. Whereas a computer is a useful tool for solving specific problems, it is no substitute for a thorough grounding in the basics of the subject.
(2) Computer software is being improved and updated all the time. The most popular and up-to-date computer package for structural engineering as I write these words may be dated (at best) or obsolete (at worst) by the time you read this. If you are interested in the latest software, look at specialist computer magazines or articles and advertisements in the civil and structural engineering and architecture press, or if you are a student, consult your lecturers.

I have set my students assignments where they have to solve a structural problem by hand then check their results by analysing the same problem using appropriate computer software. If the answers obtained by the two approaches differ, it is always instructive to find out whether the error is in the student's hand calculations (most frequently the case) or in the computer analysis (occurs less frequently, but does happen sometimes when the student has input incorrect or incomplete data – the old 'rubbish in, rubbish out'!).

The website

You will find worked solutions to some of the problems in this book at a website maintained by the publishers: www.blackwellpublishing.com/garrison. In addition, all readers can contact me via the website – your suggestions, comments and criticisms are welcome.

An overview of this book

If you are a student studying a module called Structures, Structural Mechanics or similar, the chapter headings in this book will tie in – more or less – with the lecture topics presented by your lecturer or tutor. I suggest you read each chapter of this book soon after the relevant lecture or class to reinforce your knowledge and skills in the topic concerned. I advise all readers to have a pen and paper beside them to jot down notes as they go through the book – particularly the numerical examples. In my experience, this greatly aids understanding.

- Chapters 1–5 introduce the fundamental concepts, terms and language of structures.
- Chapters 6–10 build on the basic concepts and show how they can be used, mathematically, to solve simple structural problems.
- Chapter 11 deals with the very important concept of stability and discusses how to ensure structures are stable – and recognise when they're not!
- Chapters 12–15 deal with the analysis of pin-jointed frames, a topic that some students find difficult.
- Chapter 16 covers shear force and bending moment diagrams – an extremely important topic.
- Chapters 17–20 deal with stress in its various guises.
- Structural materials are dealt with more fully in other texts, but Chapter 21 provides an introduction to this topic.
- Chapter 22 introduces structural design, which, again, is dealt with more fully in other texts.
- Chapters 23 and 24 deal, respectively, with the conceptual design of structures and the calculation of loads and will be of particular interest to students of architecture.

How to use this book

It is not necessary for all readers to read this book from cover to cover. However, the book has been designed to follow the subject matter in the order usually adopted by teachers and lecturers teaching Structures to students on degree and HND courses in Civil Engineering. If you are a student on such a course, I suggest you read the book in stages in parallel with your lectures.

- All readers should read Chapters 1–5 as these lay down the fundamentals of the subject.
- Civil engineering students should read all chapters in the book, with the possible exception of Chapters 14 and 15 if these topics are not taught on your course.
- Students of architecture should concentrate on Chapters 1–9 and 21–24, but read certain other chapters as directed by your tutor.

Let's keep it simple

James Dyson, the inventor of the dual cyclone vacuum cleaner that bears his name, discusses one of its design features – the transparent plastic cylinder within which the rubbish collects – in his autobiography:

'A journalist who came to interview me once asked, "The area where the dirt collects is transparent, thus parading all our detritus on the outside, and turning the classic design inside out. Is this some post-modernist nod to the architectural style pioneered by Richard Rodgers at the Pompidou Centre, where the air-conditioning and escalators, the very guts, are made into a self-referential design feature?"

'"No," I replied. "It's so you can see when it's full."'

(From *Against the Odds* by James Dyson and Giles Coren (Texere 2001))

It is my aim to keep this book as simple, straightforward and jargon-free as possible.

Worked solutions to the tutorial questions can be found at:
www.blackwellpublishing.com/garrison

Acknowledgements

This is my first book. My thanks go to the following people for helping to ensure it won't be my last:

- Katie Bartozzi, Amanda Brown, Andrew Brown, Fred Garrison, Jean Garrison, Simon Garrison, Pete Gordon, Paul Hirst and Phil Yates.
- Richard Hare for his assistance with some of the diagrams.
- Julia Burden and Emma Moss of Blackwell Publishing.
- My wife Jenny – my greatest fan and my fondest critic.
- Nick Crinson, whose encyclopaedic knowledge of the London Underground inspired the analogy given in Chapter 7.
- Most importantly, Jim Adams, who was the inspiration for the whole project.

And, to parrot that hackneyed catch-all used by all authors at this point, my thanks go to all the others I haven't mentioned, without whom, etc. etc. – they know who they are!

What is structural engineering?

Introduction

In this chapter you, the reader, are introduced to structures. We will discuss what a structure actually is. The professional concerned with structures is the structural engineer. We will look at the role of the structural engineer in the context of other construction professionals. We will also examine the structural requirements of a building and will review the various individual parts of a structure and the way they interrelate. Finally you will receive some direction on how to use this book depending on the course you are studying or the nature of your interest in structures.

> ### Structures in the context of everyday life
>
> There is a new confidence evident in major British cities. Redundant Victorian industrial structures are being converted to luxury apartments. Tired old 1960s shopping centres are being razed to the ground, and attractive and contemporary replacements are appearing. Public housing estates built over 40 years ago are being demolished and replaced with more suitable housing. Social shifts are occurring: young professional people are starting to live in city centres and new services such as cafés, bars and restaurants are springing up to serve them. All these new uses require new buildings or converted old buildings. Every building has to have a structure. In some of these new buildings the structure will be 'extrovert' – in other words the structural frame of the building will be clearly visible to passers-by. In many others, the structure will be concealed. But, whether seen or not, the structure is an essential part of any building. Without it, there would be no building.

What is a structure?

The structure of a building (or other object) is the part which is responsible for maintaining the shape of the building under the influence of the forces, loads and other environmental factors to which it is subjected. It is important that the structure as a whole (or any part of it) does not fall down, break or deform to an unacceptable degree when subjected to such forces or loads.

The study of structures involves the analysis of the forces and stresses occurring within a structure and the design of suitable components to cater for such forces and stresses.

As an analogy, consider the human body. Your body comprises a skeleton of 206 bones which constitutes the structure of your body. If any of those bones were to break, or if any of the joints between those bones were to disconnect or seize up, your injured body would 'fail' structurally (and cause you a great deal of pain!).

Examples of structural components (or 'members', as structural engineers call them) include:

- steel beams, columns, roof trusses and space frames;
- reinforced concrete beams, columns, slabs, retaining walls and foundations;
- timber joists, columns, glulam beams and roof trusses;
- masonry walls and columns.

Figure 1.1 shows the Lower Manhattan skyline in New York, one of the greatest concentrations of high-rise buildings in the world. Space limita-

Fig. 1.1 Lower Manhattan skyline, New York City.

tions on the island meant building construction had to proceed upwards rather than outwards, and the presence of solid rock made foundations for these soaring structures feasible.

What is an engineer?

As I mentioned in the introduction, the general public are poorly informed about what an engineer is and what he or she does. 'Engineer' is not the correct word for the man (or woman) who comes round to repair your ailing tumble drier or office photocopier – nor does it have much to do with engines! In fact, the word 'engineer' comes from the French word *ingénieur*, which refers to someone who uses his ingenuity to solve problems. An engineer, therefore, is a problem-solver.

When we buy a product – for example, a bottle-opener, a bicycle or a loaf of bread – we are really buying a solution to a problem. For instance, you would buy a car not because you wish to have a tonne of metal parked outside your house but rather because of the service it can offer you: a car solves a transportation problem. You could probably think of numerous other examples, such as:

- A can of baked beans solves a hunger problem.
- Scaffolding solves an access problem.
- Furniture polish solves a cleaning problem.
- A house or flat solves an accommodation problem.
- A university course solves an education problem.

A structural engineer solves the problem of ensuring that a building – or other structure – is adequate (in terms of strength, stability, cost, etc.) for its intended use. We shall expand on this later in the chapter. A structural engineer does not usually work alone: he is part of a team of professionals, as we shall see.

The structural engineer in the context of related professions

If I were to ask you to name some of the professionals involved in the design of buildings, the list you would come up with would probably include the following:

- the architect;
- the structural engineer;
- the quantity surveyor.

Of course, this is not an exhaustive list. There are many other professionals involved in building design (for example, building surveyors and project managers) and many more trades and professions involved in the actual construction of buildings, but for simplicity we will confine our discussion to the three named above.

The architect is responsible for the design of a building with particular regard to its appearance and environmental qualities such as light levels and noise insulation. His starting point is the client's brief. (The client usually represents the person or organisation that is paying for the work to be done.)

The structural engineer is responsible for ensuring that the building can safely withstand all the forces to which it is likely to be subjected, and that it will not deflect or crack unduly in use.

The quantity surveyor is responsible for measuring and pricing the work to be undertaken – and for keeping track of costs as the work proceeds.

So, in short:

(1) The architect makes sure the building looks good.
(2) The (structural) engineer ensures it will stand up.
(3) The quantity surveyor ensures its construction is economical.

Of course, these are simplistic definitions, but they'll do for our purposes.

Now I'm not an architect and I'm not a quantity surveyor. (My father is, but he's not writing this book.) However, I am a structural engineer and this book is about structural engineering, so in the remainder of this chapter we're going to explore the role of the structural engineer in a bit more detail.

Structural understanding

The basic function of a structure is to transmit loads from the position of application of the load to the point of support and thus to the foundations in the ground. (We'll be looking at the meaning of the word 'load' more fully in Chapter 5, but for the time being consider a load as being any force acting externally on a structure.)

Any structure must satisfy the following criteria:

(1) Aesthetics (it must look nice).
(2) Economy (it mustn't cost more than the client can afford – and less if possible).
(3) Ease of maintenance.
(4) Durability. This means that the materials used must be resistant to corrosion, spalling (pieces falling off), chemical attack, rot or insect attack.
(5) Fire resistance. While few materials can completely resist the effects of fire, it is important for a building to resist fire long enough for its occupants to be safely evacuated.

In order to ensure that a structure behaves in this way, we need to develop an understanding and awareness of how the structure works.

Safety and serviceability

There are two main requirements of any structure: it must be safe and it must be serviceable. 'Safe' means that the structure should not collapse

– either in whole or in part. 'Serviceable' means that the structure should not deform unduly under the effects of deflection, cracking or vibration.
Let's discuss these two points in more detail.

Safety

A structure must carry the expected loads without collapsing as a whole and without any part of it collapsing. Safety in this respect depends on two factors:

(1) The loading the structure is designed to carry has been correctly assessed.
(2) The strength of the materials used in the structure has not deteriorated.

From this it is evident that we need to know how to determine the load on any part of a structure. We will learn how to do this later in the book. Furthermore, we also know that materials deteriorate in time if not properly maintained: steel corrodes, concrete may spall or suffer carbonation, timber will rot. The structural engineer must consider this when designing any particular building.

Serviceability

A structure must be designed in such a way that it doesn't deflect or crack unduly in use. It is difficult or impossible to completely eliminate these things – the important thing is that the deflection and cracking are kept within certain limits. It must also be ensured that vibration does not have an adverse effect on the structure – this is particularly important in parts of buildings containing plant or machinery.

If, when you walk across the floor of a building, you feel the floor deflect or 'give' underneath your feet, it may lead you to be concerned about the integrity of the structure. Excessive deflection does not necessarily mean that the floor is about to collapse, but because it may lead to such concerns, deflection must be 'controlled'; in other words, it must be kept within certain limits. To take another example, if a lintel above a doorway deflects too much, it may cause warping of the door frame below it and, consequently, the door itself may not open or close properly.

Cracking is ugly and may or may not be indicative of a structural problem. But it may, in itself, lead to problems. For example, if cracking occurs on the outside face of a reinforced concrete wall then rain may penetrate and cause corrosion of the steel reinforcement within the concrete.

The composition of a building structure

A building structure contains various elements, the adequacy of each of which is the responsibility of the structural engineer. In this section we

Fig. 1.2 Roof structure of Quartier 206 shopping mall, Berlin.

briefly consider the form and function of each. These elements will be considered in more detail in Chapter 3.

A roof protects people and equipment in a building from weather. An example of a roof structure is shown in Fig. 1.2.

> If you plan on buying a house in the United Kingdom, be wary of buying one which has a flat roof. Some roofing systems used for waterproofing flat roofs deteriorate over time, leading to leaking and potentially expensive repairs. The same warning applies to flat-roofed additions to houses, such as porches or extensions.

Walls can have one or more of several functions. The most obvious one is *loadbearing* – in other words, supporting any walls, floors or roofs above it. But not all walls are loadbearing. Other functions of a wall include the following:

- partitioning, or dividing, rooms within a building – and thus defining their shape and extent;
- weatherproofing;
- thermal insulation – keeping heat in (or out);
- noise insulation – keeping noise out (or in);
- fire resistance;
- security and privacy;
- resisting lateral (horizontal) loads such as those due to retained earth, wind or water.

Consider the wall closest to you as you read these words. Is it likely to be loadbearing? What other functions does the wall perform?

A floor provides support for the occupants, furniture and equipment in a building. Floors on an upper level of a building are always *suspended*, which means that they span between supporting walls or beams. Ground floor slabs may sit directly on the ground beneath.

Staircases provide for vertical movement between different levels in a building. Figure 1.3 shows a concrete staircase in a multi-storey building. Unusually, the staircase is fully visible from outside the building. How is this staircase supported structurally?

Foundations represent the interface between the building's structure and the ground beneath it. A foundation transmits all the loads from a building into the ground in such a way that settlement (particularly uneven settlement) of the building is limited and failure of the underlying soil is avoided.

Fig. 1.3 A very visible staircase. How is it supported?

> On a small sandy island in the Caribbean, a low-rise hotel was being constructed as part of a larger leisure resort. The contractor for the hotel (a somewhat maverick individual) thought he could save money by not constructing foundations. He might have got away with it were it not for an alert supervising engineer, who spotted that the blockwork walls did not appear to be founded on anything more rigid than sand.
>
> A furious argument ensued between the design team and the contractor, who not only readily admitted that no foundations had been built but also asserted that, in his opinion, none was required. In a developed country the contractor would have been dismissed instantly and probably prosecuted, but things were a little more free and easy in this corner of the Caribbean.
>
> But nature exacted its own retribution. That night, a tropical storm blew up, the sea washed over the island … and the partly-built structure was entirely washed away.

In a building it is frequently necessary to support floors or walls without any interruption or division of the space below. In this case, a horizontal element called a beam will be used. A beam transmits the loads it supports to columns or walls at the beam's ends.

A column is a vertical loadbearing element which usually supports beams and/or other columns above. Laymen often call them pillars or poles or posts. Individual elements of a structure, such as beams or columns, are often referred to as *members*.

Figure 1.4 shows a conventional building enclosed in a glazed outer structure. The two structures are, apparently, completely independent of each other.

A few words for students on architecture courses

If you are studying architecture, you may be wondering why you need to study structures at all. It is not the purpose of this book to make you a fully qualified structural engineer. However, as an architect, it is important that you understand the principles of structural behaviour. Moreover, with some basic training there is no reason why architects cannot design simple structural members (e.g. timber joists supporting floors) themselves. On larger projects architects work in inter-disciplinary teams which usually include structural engineers. It is therefore important to understand the role of the structural engineer and the language and terms that the structural engineer uses.

How does the study of structures impinge on the training of an architect?

If you are on a degree course in architecture you will have formal lectures in structures throughout your course. You will also be assigned projects involving the architectural design of buildings to satisfy given require-

What is structural engineering? 9

Fig. 1.4 A conventional building enclosed in a glazed outer structure.

ments. It is essential to realise that all parts of the building need to be supported. Always ask yourself the question: 'How will my building stand up?' Remember – if you have difficulty in getting your model to stand up, it is unlikely that the real thing will stand up either!

Learn the language: a simple explanation of terms used by structural engineers

Introduction

Structural engineers use the following words (amongst others, of course) in technical discussions:

- force
- reaction
- stress
- moment.

None of these words is new to you; they are all common English words that are used in everyday speech. However, in structural engineering each of these words has a particular meaning. In this chapter we shall have a brief look at the specific meanings of the above words before exploring them in more detail in later chapters.

Force

A force is an *influence* on an object (for example, part of a building) that may cause movement. For example, the weight of people and furniture within a building causes a vertically downwards force on the floor, and wind blowing against a building causes a horizontal (or near horizontal) force on the external wall of the building.

Force is discussed more fully in Chapter 4, together with related terms such as mass and weight. Forces are also sometimes referred to as *loads* – the different types of load are reviewed in Chapter 5.

Reaction

If you stand on a floor, the weight of your body will produce a downward force into the floor. The floor reacts to this by pushing upwards with a force of the same magnitude as the downward force due to your body weight. This upward force is called a *reaction*, as its very presence is a response to the downward force of your body. Similarly, a wall or a column supporting a beam will produce an upward reaction as a response to the downward forces the beam transmits to the wall (or column) and a foundation will produce an upward reaction to the downward force in the column or wall that the foundation is supporting.

The same is true of horizontal forces and reactions. If you push horizontally against a wall, your body is applying a horizontal force to the wall – which the wall will oppose with a horizontal reaction.

The concept of a reaction is discussed in more detail in Chapter 6 and you will learn how to calculate reactions in Chapter 9.

Stress

Stress is internal pressure. A heavy vehicle parked on a road is applying pressure to the road surface – the heavier the vehicle and the smaller the contact area between the vehicle's tyres and the road, the greater the pressure. As a consequence of this pressure on the road surface, the parts of the road below the surface will experience a pressure which, because it is within an object (in this case, the road) is termed a *stress*. Because the effect of the vehicle's weight is likely to be spread, or dispersed, as it is transmitted downwards within the road structure, the stress (internal pressure at a point) will decrease the further down you go within the road's construction.

So, stress is *internal pressure* at a given point within, for example, a beam, slab or column. It is likely that the intensity of the stress will vary from point to point within the object.

Stress is a very important concept in structural engineering. In Chapters 17–20, you will learn more about how to calculate stresses.

Moment

A moment is a turning effect. When you use a spanner to tighten a nut, mechanically wind up a clock or turn the steering wheel on your car, you are applying a moment. The concept and calculation of moments is discussed in Chapter 8.

The importance of 'speaking the language' correctly

A major American bank planned changes to its London headquarters building that entailed the removal of substantial internal walls. Although a well-known firm of structural engineers was used for the design, the work itself was entrusted to a firm of shopfitters who clearly had no experience whatsoever in this type of work.

The client issued the structural engineer's drawings to the shopfitting contractor. In a site meeting, the contractor asked the structural engineer if it would be all right to use steel 'H' sections at the points where 'UC' columns were indicated on the drawings. The structural engineer was a little puzzled by this and pointed out that 'UC' stands for universal column, which are indeed steel 'H' sections. The contractor admitted, a little sheepishly, that he had thought that 'UC' stood for 'U-shaped channel section'!

The structural engineer was so shaken by this conversation and its potential consequences that he strongly advised the client to sack the shopfitters and engage a contractor who knew what he was doing.

How do structures (and parts of structures) behave?

Introduction

In this chapter we will discuss how parts of a structure behave when they are subjected to forces. We will consider the meanings of the terms *compression, tension, bending* and *shear*, with examples of each. Later in the chapter we will look at the various elements that make up a structure, and at different types of structure.

Compression

Figure 3.1 (a) shows an elevation – that is, a side-on view – of a concrete column in a building. The column is supporting beams, floor slabs and other

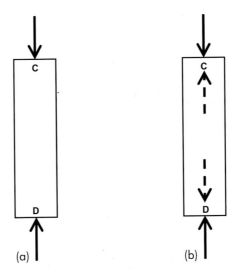

Fig. 3.1 A column in compression.

columns above and the load, or force, from all of these is acting downwards at the top of the column. This load is represented by the downward arrow at the top of the column. Intuitively, we know that the column is being squashed by this applied load – it is experiencing compression.

As we have seen briefly in Chapter 2 and will discuss more fully in Chapter 6, a downward force must be opposed by an equal upward force (or reaction) if the building is stationary – as it should be. This reaction is represented by the upward arrow at the bottom of the column in Fig. 3.1 (a). Now, not only must the rules of equilibrium (total force up = total force down) apply for the column as a whole; these rules must apply *at any and every point* within a stationary structure.

Let's consider what happens at the top of the column – specifically, point C in Fig. 3.1 (b). The downward force shown in Fig. 3.1 (a) at point C must be opposed by an upward force – also at point C. Thus there will be an upward force within the column at this point, as represented by the upward broken arrow in Fig. 3.1 (b). Now let's consider what happens at the very bottom of the column – point D in Fig. 3.1 (b). The upward force shown in Fig. 3.1 (a) at point D must be opposed by a downward force at the same point. This is represented by the downward broken arrow in Fig. 3.1 (b).

Look at the direction of the broken arrows in Fig. 3.1 (b). These arrows represent the internal forces in the column. You will notice that they are pointing away from each other. This is always the case when a structural element is in compression: the arrows used to denote compression point away from each other.

Tension

Figure 3.2 shows a heavy metal block suspended from the ceiling of a room by a piece of string. The metal block, under the effects of gravity, is pulling the string downwards, as represented by the downward arrow. The string is thus being stretched and is therefore in tension.

For equilibrium, this downward force must be opposed by an equal upward force at the point where the string is fixed to the ceiling. This opposing force is represented by an upward arrow in Fig. 3.2 (a). Note that if the ceiling wasn't strong enough to carry the weight of the metal block, or the string was improperly tied to it, the weight would come crashing to the ground and there would be no upward force (or reaction) at this point. As with the column considered above, the rules of equilibrium (total force up = total force down) must apply at any and every point within this system if it is stationary.

Let's consider what happens at the top of the string. The upward force shown in Fig. 3.2 (a) at point E must be opposed by a downward force – also at this point. Thus there will be a downward force within the string at this point, as represented by the downward broken arrow in Fig. 3.2 (b). Now let's consider what happens at the very bottom of the string – at the point where the metal block is attached (point F). The downward force shown in Fig. 3.2 (a) at point F must be opposed by an upward force at this

How do structures (and parts of structures) behave? 15

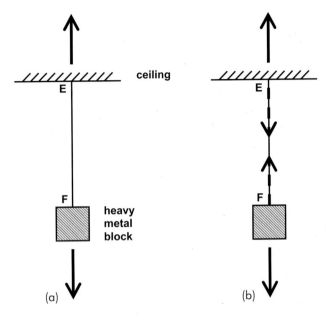

Fig. 3.2 A piece of string in tension.

point. This upward force within the string at this point is represented by the upward broken arrow in Fig. 3.2 (b).

Look at the direction of the broken arrows in Fig. 3.2 (b). These arrows represent the internal forces in the string. You will notice that they are pointing towards each other. This is always the case when a structural element is in tension: the arrows used to denote tension point towards each other. (An easy way to remember this principle is the letter T, which stands for both Towards and Tension.)

The standard arrow notations for members in (a) tension and (b) compression are shown in Fig. 3.3. You should familiarise yourself with them as we shall meet them again in later chapters.

Note: Tension and compression are both examples of *axial* forces – they act along the axis (or centre line) of the structural member concerned.

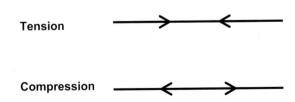

Fig. 3.3 Arrow notations for tension and compression.

Bending

Consider a simply supported beam (that is, a beam that simply rests on supports at its two ends) subjected to a central point load. The beam will tend to bend, as shown in Fig. 3.4. The extent to which the beam bends will depend on four things:

(1) The material from which the beam is made. You would expect a beam made of rubber to bend more than a concrete beam of the same dimensions under a given load.
(2) The cross-sectional characteristics of the beam. A large diameter wooden tree trunk is more difficult to bend than a thin twig spanning the same distance.
(3) The span of the beam. Anyone who has ever tried to put up bookshelves at home will know that the shelves will sag to an unacceptable degree if not supported at regular intervals. (The same applies to the hanger rail inside a wardrobe. The rail will sag noticeably under the weight of all those clothes if it is not supported centrally as well as at its ends.)
(4) The load to which the beam is subjected. The greater the load, the greater the bending. Your bookshelves will sag to a greater extent under the weight of heavy encyclopedias than they would under the weight of a few light paperback books.

If you carry on increasing the loading, the beam will eventually break. Clearly, the stronger the material, the more difficult it is to break. A timber ruler is quite easy to break by bending; a steel ruler of similar dimensions might bend quite readily but it's unlikely that you would manage to break it with your bare hands!

This is evidently one way in which a beam can fail – through excessive bending. Beams must be designed so that they do not fail in this way.

Fig. 3.4 Bending in a beam.

Shear

Consider two steel plates that overlap each other slightly, with a bolt connecting the two plates through the overlapping part, as shown in Fig. 3.5 (a). Imagine now that a force is applied to the top plate, trying to pull it to the left. An equal force is applied to the bottom plate, trying to pull it to the right. Let's now suppose that the leftward force is slowly increased, as is the rightward force. (Remember that the two forces must be equal if the whole system is to remain stationary.) If the bolt is not as strong as the plates, eventually we will reach a point when the bolt will break. After the bolt has broken, the top part of it will move off to the left with the top plate and the bottom part will move off to the right with the bottom plate.

Let's examine in detail what happens to the failure surfaces (that is, the bottom face of the top part of the bolt and the top face of the bottom part of the bolt) immediately after failure. As you can see from the 'exploded' part of Fig. 3.5 (a), the two failure surfaces are sliding past each other. This is characteristic of a shear failure.

We'll now turn our attention to a timber joist supporting the first floor of a building, as shown in Fig. 3.5 (b). Let's imagine that timber joists are supported on masonry walls and that the joists themselves support floorboards, as would be the case in a typical domestic dwelling – such as,

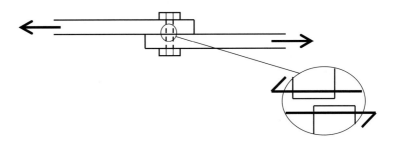

(a) Shear in a bolt connecting two plates

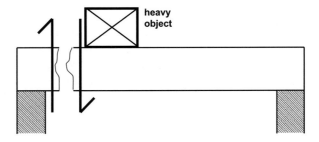

(b) Shear in a timber joist

Fig. 3.5 The concept of shear.

perhaps, the house you live in. Suppose that the joists are inappropriately undersized – in other words, they are not strong enough for the loads they are likely to have to support.

Now let's examine what would happen if a heavy object – for example, some large piece of machinery – was placed on the floor near its supports, as shown in Fig. 3.5 (b). If the heavy object is near the supporting walls, the joists may not bend unduly. However, if the object is heavy enough and the joists are weak enough, the joist may simply break. This type of failure is analogous to the bolt failure discussed above. With reference to Fig. 3.5 (b), the right-hand part of the beam will move downwards (as it crashes to the ground), while the left-hand part of the beam will stay put – in other words, it moves upwards relative to the downward-moving right-hand part of a beam. So, once again, we get a failure where the two failure surfaces are sliding past each other: a shear failure. So a shear failure can be thought of as a cutting or slicing action.

So, this is a second way in which a beam can fail – through shear. Beams must be designed so that they do not fail in this way. (Incidentally, the half-headed arrow notation shown in Fig. 3.5 is the standard symbol used to denote shear.)

The consequences of bending and shear failures – and how to design against them – will be discussed more fully in Chapter 16.

Structural elements and their behaviour

The various types of structural element that might be found in a building – or any other – structure were introduced in Chapter 1. Now we've learned about the concepts of compression, tension, bending and shear, we'll discuss how these different parts of a structure behave under load.

Beams

Beams may be ***simply-supported, continuous*** or ***cantilevered***, as illustrated in Fig. 3.6. They are subjected to bending and shear under load, and the deformations under loading are shown by broken lines.

A simply-supported beam rests on supports, usually located at each end of the beam. A continuous beam spans two or more spans in one unbroken unit; it may simply rest on its supports, but more usually it is gripped (or fixed) by columns above and below it. A cantilever beam is supported at one end only; to avoid collapse, the beam must be continuous over, or rigidly fixed at, this support.

Beams may be of timber, steel or reinforced or prestressed concrete.

Slabs

As with beams, slabs span horizontally between supports and may be simply supported, continuous or cantilevered. But unlike beams, which are

How do structures (and parts of structures) behave?   19

Fig. 3.6 Beam types.

usually narrow compared with their depth, slabs are usually wide and relatively shallow and are designed to form flooring – see Fig. 3.7.

Slabs may be one-way spanning, which means they are supported by walls on opposite sides of the slab, or two-way spanning, which means that they are supported by walls on all four sides. This description assumes that a slab is rectangular in plan, as is normally the case. Slabs are usually of reinforced concrete and in buildings they are typically 150–300 millimetres in depth. Larger than normal spans can be achieved by using ribbed or waffle slabs, as shown in Fig. 3.7 (c) and (d). Like beams, slabs experience bending.

Columns

Columns (or 'pillars' or 'posts') are vertical and support axial loads, thus they experience compression. If a column is slender or supports a non-symmetrical arrangement of beams, it will also experience bending, as shown by the broken line in Fig. 3.8 (a). Concrete or masonry columns may be of square, rectangular, circular or cruciform cross-section, as illustrated in Fig. 3.8 (b). Steel columns may be H or hollow section, as illustrated later in this chapter.

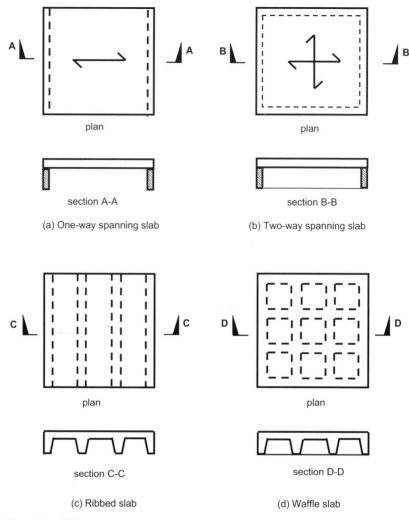

Fig. 3.7 Slab types.

Walls

Like columns, walls are vertical and are primarily subjected to compression, but they may also experience bending. Walls are usually of masonry or reinforced concrete. As well as conventional flat-faced walls you might encounter fin or diaphragm walls, as shown in Fig. 3.9. Retaining walls hold back earth or water and thus are designed to withstand bending caused by horizontal forces, as indicated by the broken line in Fig. 3.9 (c).

Foundations

As mentioned in Chapter 1, everything designed by an architect or civil or structural engineer must stand on the ground – or at least have some

How do structures (and parts of structures) behave?

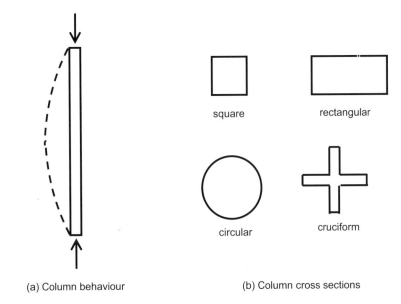

(a) Column behaviour　　　　(b) Column cross sections

Fig. 3.8 Column types.

(a) fin wall
(in plan)

(b) diaphragm wall
(in plan)

(c) retaining wall
(in cross-section)

Fig. 3.9 Wall types.

contact with the ground. So foundations are required, whose function is to transfer loads from the building safely into the ground. There are various types of foundation. A *strip* foundation provides a continuous support to loadbearing external walls. A *pad* foundation provides a load-spreading support to a column. A *raft* foundation takes up the whole plan area under a building and is used in situations where the alternative would be a large number of strip and/or pad foundations in a relatively small space. Where the ground has low strength and/or the building is very heavy, *piled* foundations are used. These are columns in the ground which transmit the

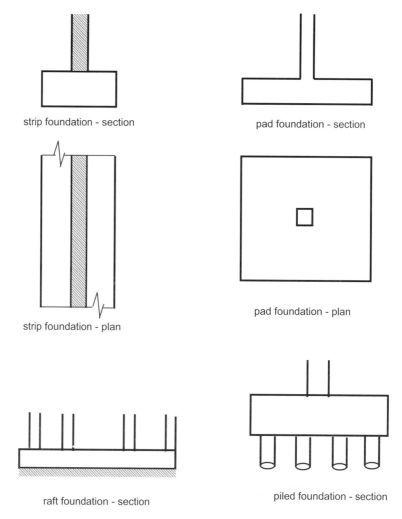

Fig. 3.10 Foundation types.

building's loads safely to a stronger stratum. All these foundation types are illustrated in Fig. 3.10.

Foundations of all types are usually of concrete, but occasionally steel or timber may be used for piles.

Arches

The main virtue of an arch, from a structural engineering point of view, is that it is in compression throughout. This means that materials that are weak in tension – for example, masonry – may be used to span considerable distances. Arches transmit large horizontal thrusts into their supports, unless horizontal ties are used at the base of the arch. It is to cope with

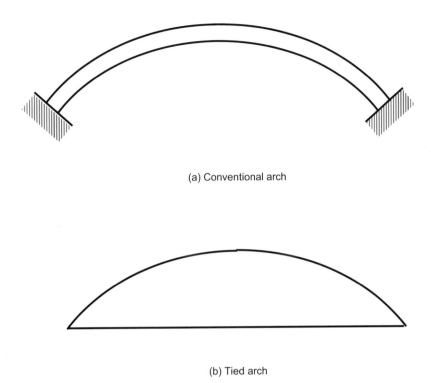

(a) Conventional arch

(b) Tied arch

Fig. 3.11 Arch types.

these horizontal thrusts that flying buttresses are provided in medieval cathedrals – see Fig. 3.11.

Trusses

A truss is a two- or three-dimensional framework and is designed on the basis that each 'member' or component of the framework is in either pure tension or pure compression and does not experience bending. Trusses are often used in pitched roof construction: timber tends to be used for domestic construction and steel caters for the larger roof spans required in industrial or commercial buildings. Lattice girders, which are used instead of solid deep beams for long spans, work on the same principle – see Fig. 3.12.

Portal frames

A portal frame is a rigid framework comprising two columns supporting rafters. The rafters may be horizontal or, more usually, inclined to support a pitched roof. Portal frames are usually of steel but may be of precast concrete. They are usually used in large single-storey structures such as warehouses or out-of-town retail sheds – see Fig. 3.13.

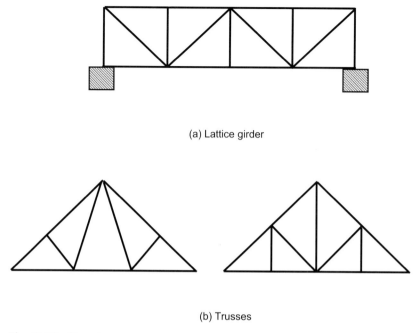

(a) Lattice girder

(b) Trusses

Fig. 3.12 Truss types.

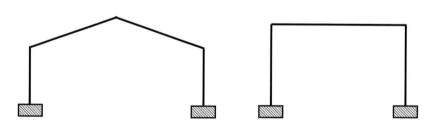

Fig. 3.13 Portal frame types.

Cable stayed and suspension structures

Cable stayed structures are usually bridges but are sometimes used in building structures where exceptionally long spans are required. Instead of being supported from below by columns or walls, the span is supported from above at certain points by cables which pass over supporting vertical masts and horizontal outriggers to a point in the ground where they are firmly anchored. The cables are in tension and must be designed to sustain considerable tensile forces – see Fig. 3.14.

Figure 3.15 shows New York's Brooklyn Bridge. Conceived by John Roebling and completed by his son Washington in 1883, the Brooklyn Bridge was the first suspension bridge in the world to use steel for its main cables

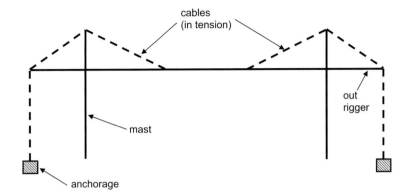

(a) Cable-stayed structure (in cross-section)

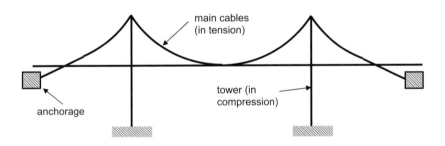

(b) Suspension bridge

Fig. 3.14 Cable stayed and suspension structures.

and was the longest suspension bridge in the world at the time of its construction. The foundations were excavated from within underwater caissons using compressed air, causing crippling illnesses among the workers, including Washington Roebling himself.

Cross-sectional types

There is an infinite range of cross-sectional shapes available. Standard sections are illustrated in Fig. 3.16.

- Beams and slabs in timber and concrete are usually rectangular in cross-section.
- Concrete columns are usually of circular, square, rectangular or cruciform cross-section (see above).

Fig. 3.15 Brooklyn Bridge, New York City.

- Steel beams are usually of 'I' or hollow section.
- Steel columns are usually of 'H' or hollow section.
- Prestressed concrete beams are sometimes of 'T', 'U' or inverted 'U' section.
- Members of steel trusses are sometimes of channel or angle sections.
- Steel Z purlins (not illustrated) are often used to support steel roofing or cladding.

Appraisal of existing structures

> ### Steam room indiscretion
>
> One Saturday morning I was relaxing in the steam room at my local fitness centre after a punishing workout. My companions there were two men in their 20s and an older man, possibly mid-40s. All were sat in their swimming trunks, happily sweating away.
>
> The two younger men were clearly friends. As I entered the steam room, they were in the middle of a conversation about a forthcoming party, which went something like this:
>
> MAN 1: Is Craig going?
> MAN 2: Yes, I think so.
> MAN 1: Oh, good – he's a real laugh.
> MAN 2: Yes – he's a nutcase. (Pause) Craig was leaving the rugby club last Sunday lunchtime after a few drinks when he smashed his car up. So he pushed it off the road into a field, covered it with straw, then phoned the police and reported it stolen.

> MAN 1: (After a slight pause) Did he get away with it?
> MAN 2: Er – yes, I think so.
>
> The older man, who'd remained silent up to this point, now spoke up slowly and deliberately: 'The thing that you've got to remember is that when you're in a steam room, police officers are not wearing uniforms.'
>
> There was an uncomfortable silence while the significance of this remark sunk in. The police officer eventually smoothed things over by telling a similar anecdote of his own, after which he left the steam room. One of the young men turned to me and his friend and said: 'Well, that could have been a bit unfortunate, couldn't it!'
>
> 'Yes,' I replied, 'I'm CID.'

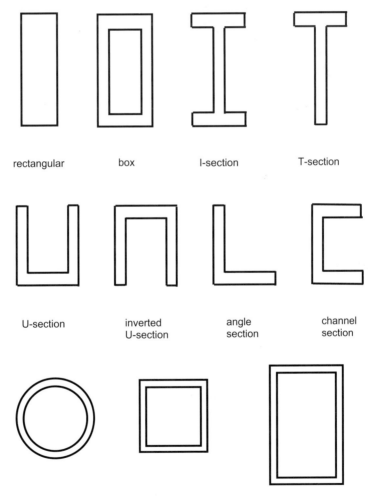

Fig. 3.16 Cross-section types.

The story above has a serious message. People are not always what they seem: a near-naked police officer looks much the same as anyone else. Similarly, structures are not always what they might seem either, although the problem here is usually one of too much cladding. In some buildings the designers choose to make a feature of the structure; in others, the structure is totally concealed.

In your future professional career you may be called upon to carry out a structural inspection of an existing building, usually after someone else – perhaps a building surveyor – has identified a fault that he suspects may be structural in nature. It is not always easy to assess how an existing building functions structurally. Certainly, you can pick up clues from the age and style of the building, and original drawings of the structure as built are very useful – in the unlikely event that they are available.

So, if you have to carry out a structural appraisal of an existing building, my advice is: tread carefully.

What you should remember from this chapter

The concepts of compression, tension, bending and shear are fundamental to any study of structural mechanics. The reader should clearly understand the meaning and implications of each. Different elements of a structure deform in different ways under load. The reader should understand and be able to visualise these patterns of structural behaviour, which are fundamental to structural design.

Force, mass and weight

Introduction

In this chapter we look at force, mass and weight – their definitions, the relationships between them, their units of measurement and their practical application.

Force

We use the term *force* in everyday life. For example, somebody may force you to do something. This means that that person, through their words, actions or other behaviour, compels you to take a certain course of action. The word force in a technical context is similar: a force is an influence, or action, on a body or object which causes – or attempts to cause – movement. For example:

- The man shown in Fig. 4.1 is pushing against a wall. In doing so, he is applying a horizontal force to that wall – in other words, he is attempting to push that wall away from him.
- Figure 4.2 shows a man standing on a hard surface. The weight of his body is applying a vertical (downwards) force on the floor – in other words, he is attempting to move the floor downwards.

Force is measured in units of Newtons (N) or kiloNewtons (kN) – but more of that later.

Mass

Mass is the amount of *matter* in a body or object. It is measured in units of grams (g) or kilograms (kg). Mass should not be confused with weight (see below).

Fig. 4.1 Man leaning against a wall.

Weight

If you have studied physics, or science generally, you will have come across the following equation:

Force = Mass × Acceleration

A much more useful form of this equation to engineers is:

Weight = Mass × Acceleration due to Gravity

If an object is dropped from a great height, it will accelerate – that is, consistently increase in speed – as it heads towards the ground. This acceleration is called the Acceleration due to Gravity and its value on this planet is 9.81 metres/sec^2. This means that a dropped object will fall 9.81 metres in the first second, (9.81 + 9.81) = 19.62 metres in the second second and (9.81 + 9.81 + 9.81) = 29.43 metres in the third second. The mass of the object is irrelevant in this context – a bundle of feathers falls at the same rate as a large lump of lead, as they experience the same rate of acceleration.

Fig. 4.2 Man standing on a floor.

Looking at the above equation, we see that the relationship between mass and weight is governed by the acceleration due to gravity. It suggests that an object of a given mass will weigh less on a planet where the gravitational pull is less. If you watch television footage of the Apollo moon landings in the late 1960s, you will notice that the astronauts appear to be leaping and bounding around on the moon in a manner that would be regarded as undignified on earth. This is because although the mass of a particular astronaut (that is, the amount of matter in his body) is obviously the same on the moon as it is on earth, the gravitational pull (and hence acceleration due to gravity) is much less on the moon and therefore the astronaut's weight is much less on the moon than it is on earth.

The relationship between weight and mass

Coming back to earth, if an object of mass 1 kg is subjected to the acceleration due to gravity of 9.81 m/sec^2 (which is approximately 10 m/sec^2), then the above equation tells us that the object's weight is (1 × 10) = 10 N. Note that weight is a force and is measured in the same units as force: Newtons (N) or kiloNewtons (kN). So:

A *weight* of 10 N is equivalent to a *mass* of 1 kg

As you are probably aware, the kilo- prefix means '1000 times', so:

1,000 N = 1 kN

Therefore, if a weight of 10 N is equivalent to a mass of 1 kg, then a weight of 1000 N (or 1 kN) is equivalent to a mass of 100 kg.

One further relationship: a weight of 10 kN is known as a tonne (also known as a metric tonne).

What do these units signify in everyday terms?

(1) Sugar is sold in your local supermarket in 1 kg bags. If you lift a 1 kg bag of sugar, you will get some idea of what a mass of 1 kg (and hence a force of 10 N) feels like.
(2) 1 kN is equivalent to 100 kg, which in turn is approximately 220 pounds or just under 16 stone – 16 stone is the weight of a reasonably large man. If you imagine a large man or woman of your acquaintance, then the mass of their body is imposing a 1 kN force downwards.
(3) A small modern car weighs about a tonne (or 10 kN).

To summarise:

- 10 N = 1 kg (a bag of sugar)
- 1000 N = 1 kN (a 16-stone person)
- Therefore: 1 kN = 100 kg
- 1000 kg = 10 kN = 1 tonne (a small car)

Density and unit weight

The *density* of a material can be calculated as follows:

$$\text{Density (kg/m}^3\text{)} = \frac{\text{Mass (kg)}}{\text{Volume (m}^3\text{)}}$$

Unit weight is a similar concept to density. The unit weight is the weight of a material per unit volume and is measured in kN/m³. Unit weights of some common building materials are given in Appendix 1.

Units generally

You should always be conscious of the units you are using in any structural calculation. Incorrect use and understanding of units can lead to wildly inaccurate answers.

The lecturers and tutors who mark and assess your coursework and examinations are well aware of the perils of getting the units wrong. Make sure that you express the units in any written work you do. For example, the force in a column is not 340, but it might be 340 kN. Omitting units is sheer laziness and it may lead the person assessing your work to doubt whether you understand what you're doing and he or she will mark the work accordingly.

Relationships with other measuring systems

Although the metric system is now generally used in scientific and technical work in the United Kingdom, you will need to know how to convert from the Imperial system of measurement (pounds, feet, inches, etc.). This is because, in your future professional career:

(1) You may have to review calculations or drawings made before the 1960s when the metric system came into use.
(2) You may be working in (or for) a country which doesn't use the metric system.
(3) You may be dealing with a profession that feels more comfortable with non-metric units.

For example:

- 1 pound = 0.454 kg
- 1 inch = 25.4 millimetres

For a more comprehensive list of conversions between different systems of measurement, see Appendix 2.

What you should remember from this chapter

- Mass is the amount of matter in an object and is measured in grams (g) or kilograms (kg).
- Weight is a force and is measured in Newtons (N) or kiloNewtons (kN).
- Density is the ratio of mass to volume and is measured in kg/m^3.
- Unit weight is the weight of a material per unit volume and is measured in kN/m^3.
- In any calculations in structures the units used should always be expressed.

Tutorial examples

Answers are given at the end of the chapter.

(1) Calculate the weight, in kN, of each of the following two people:
  (a) A young woman with a mass of 70 kg.
  (b) A middle-aged man with a mass of 95 kg.
  What would be the weights of each of these people on the moon if the gravitational acceleration on the moon is one-sixth of that on earth?

(2) Calculate the mass of a brick of length 215 mm, breadth 102.5 mm and height 65 mm if its density is 1800 kg/m³. What would be the weight of this brick?

(3) Calculate the weight of a 9 metre long reinforced concrete beam of breadth 200 mm and depth 350 mm if the unit weight of reinforced concrete is 24 kN/m³.

(4) As we will see in later chapters, the term *live load* is used to describe non-permanent load within a building – that is, those loads due to people and furniture. If a university classroom is 12 metres long and 10 metres wide and is designed to accommodate up to 60 students, calculate the live load in the classroom when full. (Note that you will have to make an assessment of the weight of an individual student, desk and chair.) Compare your answer with the British Standard value of live load (3.0 kN/m²) for classrooms.

(5) An international hotel chain plans to upgrade its hotel in a particular glamorous and exotic location by installing a rooftop swimming pool on top of its existing high-rise bedroom block. The swimming pool will be 25 metres long and 10 metres wide and will vary uniformly in depth from 1 metre to 2 metres. Calculate the volume of water in the pool. If the unit weight of water is 10 kN/m³, calculate the weight of water in the pool, in tonnes. If a small modern car weighs 1 tonne, calculate the number of cars that would be equivalent, in weight, to the water in the proposed swimming pool. If you were appointed as structural engineer for the project, what would be your initial advice to the architect and client?

(6) You are involved in a housing development project. You measure the site on a plan and find that it is rectangular, of length 300 metres and width 250 metres. 'What's the area in acres?' the developer asks you. What is your reply? (Hint: refer to Appendix 2.)

(7) You were delighted to win the full £1 million prize money during your recent appearance on a television quiz programme. However, your elation abates somewhat when the programme's producer informs you that the prize money will be given to you in cash, entirely in pound coins. Calculate the mass, weight and volume of 1 million pound coins.

Bearing in mind the sudden interest shown in you by several tabloid newspapers, explain how you would transport the cash from the television studio to your home 200 miles away.

Noting your concerns, the producer offers to pay your prize money in £2 coins instead. Calculate the appropriate mass, weight and volume for this case. Would you accept or decline this offer?

On your eventual arrival at home with your haul, you decide to store the money in a spare bedroom. Assuming conventional timber joist floor construction, do you think this would pose a structural problem and why?

Tutorial answers

(1) (a) 0.7 kN; (b) 0.95 kN. On moon: (a) 0.117 kN; (b) 0.158 kN.
(2) 2.52 kg; 0.025 kN.
(3) 15.1 kN.
(4) Your answer will probably be in the range 0.5–1.0 kN/m², depending on your assumptions.
(5) 375 m³; 375 tonnes.
(6) 18.5 acres.

Loading – dead or alive

Introduction: what is a load?

As discussed in Chapter 2, a *load* is a force on a part of a structure. The term 'load' is frequently used in everyday life. We refer to 'loading' a washing machine when we fill it with clothes to be washed. You 'load up' your car before going on a motoring holiday. The airline industry uses the term 'load factors' when describing how many passengers get on flights, and an insurance company will 'load' your premium (in other words, increase the amount you have to pay for your insurance policy) if you give information which it feels increases the risk of it having to pay out. In short, you are already familiar with the word load and its use in structural engineering is, I hope, easy to understand.

In structures, there are the following different types of loading:

- Dead load (or permanent load): as its alternative name of *permanent load* suggests, a dead load is always present. Examples of dead load include the loads – or forces – due to the weights of the various elements of construction, such as floors, walls, roofs, cladding and permanent partitions. These items – and their weights – are obviously always there, 24 hours a day, 365 days a year.
- Live load (or imposed load): live loads are not always present. They are produced by the *occupancy* of the building. Examples of live load include people and furniture. Other examples include snow loads on roofs. By their very nature, live loads, unlike dead loads, are variable. For example, a 300-seat cinema auditorium would be full of people on a Saturday evening if a major new blockbuster movie was being shown, but it might be only a quarter full on a weekday afternoon. And, of course, it would be empty when the cinema is closed. So the live load in this cinema auditorium is represented by anything between 0 and 300 people. A classroom in a college or university is a similar example. The classroom might be full of students or empty – or anything in be-

tween. Also, it might be decided to temporarily remove the desks and chairs from the classroom – to hold an exhibition there, for example.

Because of this variability, live loads are treated differently to dead loads in structural design.

A live load does not have to be moving, animate or alive in any way. For example, a dead body in a mortuary is a live load because it is there only temporarily. A car in a multi-storey car park will be a live load whether or not it is moving; again, the assumption is that the car is there for only a certain period of time and then it will be removed.

In short, dead loads are there all the time, live loads are not.

- Wind load (an example of lateral loading): **u**nlike dead and live loads, which are usually vertical in direction, wind loads act horizontally or at a shallow angle to the horizontal. Wind loads vary across the country and across the world and their effects vary according to the type of physical environment (city centre, suburban, open moorland, etc.) and the height of the building. Wind loads can act in any plan direction and their intensity can vary continually.

You can no doubt imagine that the effects of wind loads were particularly significant in the design of the London Eye and the Millennium Bridge, shown in Figs 5.1 and 5.2 respectively. Built to celebrate the millennium, Marks Barfield Architects' London Eye is a steel and glass ferris wheel structure resembling a giant bicycle wheel. Many structural and logistical problems had to be overcome in its design and construction. The Millennium Bridge is a low-slung suspension bridge and is infamous for its vibration problems – now solved – when it first opened.

Fig. 5.1 London Eye.

Fig. 5.2 Millennium Bridge, London.

Other loads

Lateral (or horizontal) loads other than wind loads include those due to earth pressures (on retaining walls, for example) or water pressures (on the side walls of water tanks). Other loads may include those due to earthquakes or subsidence.

Why do we differentiate between these types of loading?

Because the various loads described above differ in nature, we have to handle them in differing ways when we undertake structural design. For example, the total dead load in a given building remains constant unless building alterations are carried out, but live load can vary on an hour-by-hour basis. We will revisit this when we consider the basics of structural design in Chapter 22.

Nature of load

As well as considering the different types of loading we have to consider the nature of loads. This could be one of three types:

(1) Point load
(2) Uniformly distributed load
(3) Uniformly varying load.

Let's consider each of these in turn – refer to Fig. 5.3.

Loading – dead or alive   39

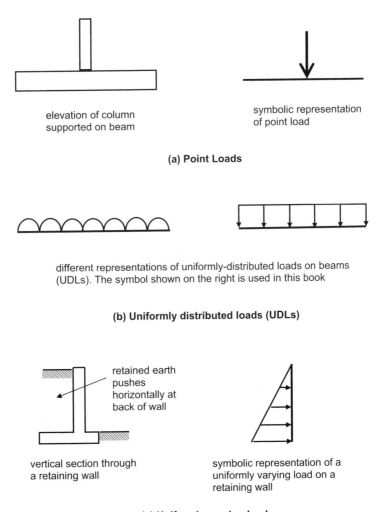

Fig. 5.3 The nature of loading.

Point load

This is a load that acts at a single point. It is sometimes called a *concentrated load*. An example would be a column supported on a beam. As the contact area of the column on the beam would be small, the load is assumed to be concentrated at a point. Point loads are expressed in units of kN and are represented by a large arrow in the direction that the load or force acts, as shown in Fig. 5.3 (a).

Uniformly distributed load

A uniformly distributed load (often abbreviated to UDL) is a load that is evenly spread along a length or across an area. For example, the loads supported by a typical beam – the beam's own weight, the weight of the floor slab it's supporting and the live load supported by the floor slab – are consistent all the way along the beam. UDLs along a beam (or any other element that is linear in nature) are expressed in units of kN/m. Similarly, the loads supported by a slab will be consistent across the slab and because a slab has area rather than linear length, UDLs on a slab are expressed in units of kN/m^2. There are at least two different symbols used for UDL, as shown in Fig. 5.3 (b).

Uniformly varying load

A uniformly varying load is a load that is distributed along the length of a linear element such as a beam, but instead of the load being evenly spread (as with a UDL) it varies in a linear fashion. A common example of this is a retaining wall. A retaining wall is designed to hold back earth, which exerts a horizontal force on the back of the retaining wall. The horizontal force on the retaining wall becomes greater the further down the wall you go. Thus the force will be zero at the top of the retaining wall but will increase linearly to a maximum value at the bottom of the wall – see Fig. 5.3 (c).

Load paths

It is important to be able to identify the *paths* that loads take through a building. As an example, we will consider a typical steel-framed structure, which comprises vertical columns arranged on a grid pattern, as shown in Fig. 5.4 (a). At each level of the building the columns will support beams, which span between the columns. Each beam may support secondary beams, which will span between the main (primary) beams. The beams will support floor slabs, usually of reinforced concrete or a steel/concrete composite construction. The floor slabs support their own weight and the live loads on them. Figure 5.4 (b) shows a typical part of the structural floor plan. A, C, D and F are columns. The lines AC, AF, DF and CD represent primary beams and line BE represents a secondary beam. A concrete floor slab spans between beams AF and BE. Another concrete floor slab spans between beams BE and CD.

Assume a heavy piece of equipment is located at point G. Clearly, it is supported by the concrete floor slab beneath it. The concrete floor slab, in turn, is supported by beams AF and BE, so it follows that the equipment is supported by those beams as well. As point G is closer to BE than AF, it can be deduced that beam BE takes a greater share of the equipment load than does beam AF.

Loading – dead or alive   41

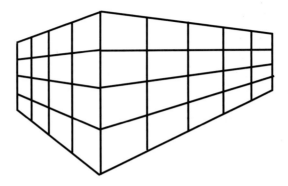

(a) typical steel framed structure

(b) part of structural floor plan   (c) load paths added

Fig. 5.4   Load paths in a structure.

Beam BE in turn is supported by primary beams AC and DF. As point G is closer to DF than AC, beam DF will support a greater share of the equipment load than beam AC. (We'll find out how to calculate these 'shares' when we look at *reactions* later in the book.)

Columns D and F support the two ends of beam FD. As beam BE sits exactly half way along DF, it will inflict a point load at the midpoint of DF, which will be shared equally between the two supporting columns D and F. Additionally, column F will support a portion of the equipment load transmitted via beam AF. Similarly, columns A and C will also take a share of the equipment load, via beam AC. Additionally, column A will support a portion of the equipment load transmitted via beam AF.

The broken arrows in Fig. 5.4 (c) indicate the paths taken through the structure by the equipment load at point G. The columns will transmit the equipment loads – along with all the other loads in the structure, of course

– down to the foundations, which will need to be strong enough to safely transmit the loads into the ground below.

To simplify the explanation, we've considered only the load paths due to one particular load. Of course, there are many loads in a building and all of them need to be considered in structural analysis and design. We will look at some examples of this in Chapter 24.

> It's one thing for a designer to impose a weight limit on a bridge; it's quite another thing to ensure that the weight limit is actually observed.
>
> A marine leisure complex was being constructed on a small tropical island adjacent to a glamorous holiday location. A concrete bridge was designed to link the island with the mainland. As the bridge would be carrying only 'trams' (the sort of trolley-like vehicles common in theme parks) which would transport visitors to the attraction from the car park on the mainland, the bridge was designed for a vehicle weighing 4 tons.
>
> During construction of the leisure complex itself it was observed that 25-ton fully laden concrete wagons were crossing the bridge. There were no signs indicating a weight limit and there were no physical barriers to stop anyone crossing. Fortunately, the bridge was clearly grossly over-designed and did not fail – or even show any signs of distress – under these loads.

Equilibrium – a balanced approach

Introduction

The eminent scientist Isaac Newton (1642–1727) is perhaps best known for his three Laws of Motion. If you have studied physics you will have come across these before. One of them gives the *force = mass × acceleration* formula mentioned in Chapter 4.

In this chapter we are concerned with Newton's Third Law of Motion, which essentially states:

'For every action there is an equal and opposite reaction.'

This means that if an object is stationary – as a building, or any part of it, usually is – then any force on it must be opposed by another force, equal in magnitude but opposite in direction. In other words, a condition of *equilibrium* will be established.

> Figure 6.1 shows a steel arch bridge, a small-scale version of the Tyne Bridge in Newcastle and the Sydney Harbour Bridge in Australia. The nature of the forces within arches leads to horizontal outward forces being generated at the ends of the arch. For equilibrium to occur, these outward thrusts must be opposed by inward (i.e. opposite) forces. These inward forces might be in the form of the reaction of a solid abutment to the bridge. Alternatively, as we see with this bridge, the road deck acts as a horizontal tension member which ties the two ends of the arch to each other, thus catering for the outward thrusting forces.

Fig. 6.1 Steel arch bridge.

Vertical equilibrium

Vertical equilibrium dictates that:

Total force upwards = Total force downwards

For example, if a man weighing about 16 stone stands on the floor of a room, the downward force into the floor due to the weight of his body is 1 kN. Assuming that the floor is stationary, it must be pushing up (or reacting) with an upward force of 1 kN.

Let's consider what would happen if the floor did not react with the same upward force as the downward force encountered. If the floor could, for some reason, muster an upward force of only, say, 0.5 kN in response to the man's downward force of 1 kN, the floor would not be capable of supporting the man's weight. The floor would break and the man would fall through it. On the other hand, if the floor were to react to the man's weight by supplying an upward force of, say, 2 kN, the man would go flying through the air like a human cannonball.

In each of the above two cases, we can see that if the upward and downward forces don't balance, movement occurs (either upward or downward). If neither of these things is occurring (in other words, the man is neither falling through the floor nor shooting through the air), we can conclude that, because everything is stationary, the forces are balanced and vertical equilibrium is observed.

Horizontal equilibrium

This tells us that:

Total force to the left = Total force to the right

Example 1: 'Tug of war'

As you may know, a tug of war is a physical competition involving two teams and a very long piece of rope. The two teams normally comprise equal numbers of contestants. Each team distributes itself along one end of the rope, as illustrated in Fig. 6.2. The team at the left end of the rope is using all its strength to pull the rope (and the opposing team) to the left. Similarly, the team at the right end of the rope is using all its strength to pull the rope to the right. If there is a river separating the two teams, the stronger team will eventually win by pulling the opposing team into the river.

A marker flag is fixed to the midpoint of the rope. Suppose you are an adjudicator, watching the competition's progress from a distance. You will be watching the marker's position. If the flag starts to move to the left, you will interpret this as meaning that the left-hand team is winning. In other words, the force to the left is greater than the force to the right. As Newton's Third Law tells us, movement occurs because the two forces are unbalanced. Similarly, if the flag starts moving to the right, this would indicate that the right-hand team is winning – because the force to the right is greater than the force to the left. Again, the two forces are unbalanced, causing movement.

However, if the flag doesn't move at all but stays in exactly the same position no matter how hard the two teams strain and pull, you would deduce that the two teams are evenly matched and neither is winning. In this case, the flag doesn't move because the force to the left is exactly the same as the force to the right. In other words, if the marker on the rope in a tug of war – or any other object – is stationary, then the force to the left and the force to the right are the same. So we have horizontal equilibrium.

Fig. 6.2 Tug of war.

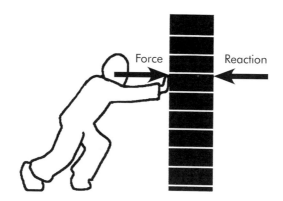

Fig. 6.3 Pushing against a wall.

Example 2: The leaner

If you lean against a wall, as shown in Fig. 6.3, your body is applying a horizontal force to the wall – to the right in the case shown in Fig. 6.3. The wall will react by providing a force (or reaction) to the left, equal in magnitude to the force applied.

If, for some reason, the wall is not able to provide an equal and opposite horizontal reaction, it means that either the wall is not strong enough or it's not fixed properly to the floor. In either case the wall will yield and movement will occur.

What we've learned about horizontal and vertical equilibrium is summarised in Fig. 6.4.

The application of equilibrium

As buildings are usually stationary, Newton's Third Law tells us that the forces on a building – or any part of it – must be in equilibrium.

Consider the beam shown in Fig. 6.5. The beam is supported on columns at each of its two ends and supports vertical loads $F1$, $F2$ and $F3$ at various points along its length. Where there are downward forces, there must be opposing upward forces, or reactions. (The term 'reaction' was introduced in Chapter 2.) Let's call the reaction at the left-hand end of the beam $R1$. The reaction at the right-hand end of the beam we will call $R2$. Using our knowledge of vertical equilibrium we can say:

Total force up = Total force down

So:

$R1 + R2 = F1 + F2 + F3$

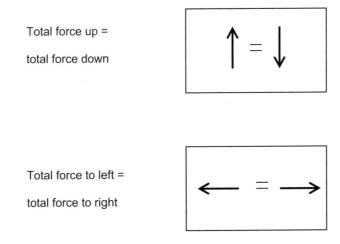

Fig. 6.4 Equilibrium.

Now if $F1 = 5$ kN, $F2 = 10$ kN and $F3 = 15$ kN, then:

$R1 + R2 = 5 + 10 + 15$ kN

So

$R1 + R2 = 30$ kN

It would be useful to calculate $R1$ and $R2$, as they represent the forces in the supporting columns. But the equation above doesn't tell us what $R1$ is and it doesn't tell us what $R2$ is; it merely tells us that the sum of the two is 30 kN. In order to evaluate each of $R1$ and $R2$, we need to know more. We will continue this theme in Chapter 9.

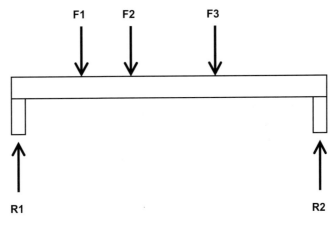

Fig. 6.5 Application of vertical equilibrium.

> The late Yorkshire veterinary surgeon James Herriot describes an encounter with a bull in a confined space in his book *Vet in Harness* (Michael Joseph, 1974). The bull, which had just received an injection from Mr Herriot, decided to lean against the vet, thus sandwiching him between its body and a wooden partition. As we know from Chapter 3, this action would have put Mr Herriot's body into compression, held in place by opposing reactions from the bull and the wooden partition. As he puts it: 'I was having the life crushed out of me. Pop-eyed, groaning, scarcely able to breathe, I struggled with everything I had, but I couldn't move an inch. ... I was certain my internal organs were being steadily ground to pulp and as I thrashed around in complete panic the huge animal leaned even more heavily.
>
> 'I don't like to think what would have happened if the wood behind me had not been old and rotten, but just as I felt my senses leaving me there was a cracking and splintering and I fell through into the next stall.'
>
> So suddenly – and fortunately for Mr Herriot – the force from the bull overcame the wooden partition's strength and it collapsed. It was no longer able to provide a reaction to keep Mr Herriot in compression and thus probably saved his life.

What you should remember from this chapter

If any object (for example, a building or part of one) is stationary then it is in equilibrium. This means that the forces on it must balance, as follows:

- Total force upwards = Total force downwards
- Total force to the left = Total force to the right

7 More about forces: resultants and components

Introduction

In previous chapters you have learned what a force is. In this chapter we will look at how forces – individually or in groups – may be handled. You will learn how to combine forces into *resultants* and how to 'split' forces into *components*.

Let's start by considering an analogy.

The Underground analogy

Imagine that you are in London and are planning a journey on the Underground railway system there. You are at Green Park station and want to travel to Oxford Circus. You consult the diagram of lines and stations displayed at the station entrance, a representation of the relevant part of which is shown in Fig. 7.1. You work out that the quickest way to reach Oxford Circus from Green Park is to travel directly there on the Victoria Line. Oxford Circus is only one station from Green Park.

However, as you enter the station, you pass a blackboard on which has been written: 'Victoria Line Closed due to Technical Difficulties.' Clearly, this news means you must change your travel plans. Assuming that you don't now decide to walk or take a bus or taxi, there are two options available to you if you wish to reach Oxford Circus as quickly as possible:

(1) Take the Jubilee Line northwards to the next station, Bond Street, where you can change onto the eastbound Central Line and travel to the next station, Oxford Circus.
(2) Take the Piccadilly Line eastwards to the next station, Piccadilly Circus, where you can change onto the northbound Bakerloo Line and travel to the next station, Oxford Circus.

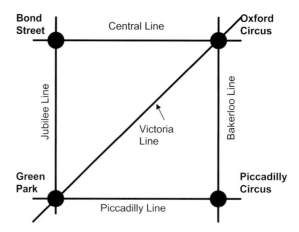

Fig. 7.1 Representation of part of London Underground network.

Clearly one of these two options will be quicker than the other, depending on frequency of trains and the ease of transferring between platforms at the interchange station. Although it is difficult to predict which option would deliver you to Oxford Circus more quickly, we can say with confidence that either route will take you – eventually – to Oxford Circus.

If we represent a journey by an arrow in the direction of the journey – with the length of the arrow representing the length of the journey – our two route options can be illustrated by the two diagrams in Fig. 7.2. In each case, the desired direct route (on the temporarily unavailable Victoria Line) has been indicated by a broken arrow. As expected, each indirect route is longer (in distance) than the direct route between Green Park and

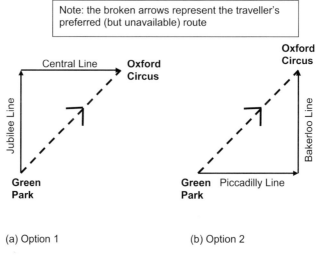

Fig. 7.2 Green Park to Oxford Circus: route options.

Oxford Circus stations, but the end result is the same: in each case, you end up at Oxford Circus station.

Whichever option you choose, your starting point is Green Park station and your finishing point is Oxford Circus station.

We will return to this analogy later in the chapter.

Resolution of forces

We encountered the concept of a force in Chapter 4. As stated there, a force is an influence or action on a body that causes – or attempts to cause – movement. Forces can act in any direction, but the direction in which a given force acts is important. You will know from your studies of mathematics that something that has both magnitude and direction is called a *vector* quantity. As force has both magnitude and direction, force is an example of a vector quantity.

To define a given force fully, we need to state its:

(1) magnitude (for example, 50 kN)
(2) direction, or line of action (for example, vertical)
(3) point of application (for example, 2 metres from the left-hand end of a beam).

What happens when several forces act at the same point?

Clearly it is possible that several forces may act at the same point. These forces may all be different in magnitude and acting in different directions. It would be convenient if we could simplify these forces in such a way that they are represented by just one force, acting in a certain direction. This one force is called the *resultant* force.

The 'Donald and Tristan' analogy

Consider a trolley standing in the middle of a large room with a highly polished wooden floor. The trolley is a piece of furniture with wheels or castors that make it easy to push it in any direction – rather like the sweet trolleys used in expensive restaurants. The room is otherwise empty. Donald enters the room and starts to push the trolley in an easterly direction. The trolley moves eastwards, as shown in Fig. 7.3 (a). At that moment, Donald's friend Tristan enters the room and starts to push the trolley northwards, while Donald continues to try to push the trolley eastwards.

As you would expect, under the influence of the two friends pushing the trolley in different directions, the trolley now moves off in a generally north-easterly direction. Figure 7.3 (b) indicates this activity, as viewed from above (a plan view), with the broken arrow representing the movement of the trolley. But in what direction, exactly, would the trolley move?

Well, it depends on the relative effort Donald and Tristan put into the exercise. If Donald puts a lot of energy into his easterly push, while Tristan

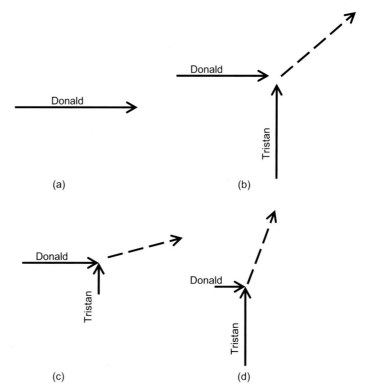

Fig. 7.3 Resultants of forces.

makes a puny attempt at his northbound shove, the trolley will move off in the direction shown by the broken arrow in Fig. 7.3 (c). (The length of the arrows represents the magnitude of the forces.) On the other hand, if Tristan exerts himself fully with his northerly push and Donald can't be bothered to put much effort into his east-bound push, the trolley will move off in the direction shown by the broken arrow in Fig. 7.3 (d). (Again, the length of the arrows represents the magnitude of the forces.) In each case there are two forces involved: one from Donald, the other from Tristan. As we have seen, these two forces are of different magnitudes and act in different directions.

In each of Figs 7.3 (b)–(d) the broken arrow represents the *resultant* force. In each case, the magnitude and direction of this resultant force represents the combined effect of Donald and Tristan's pushing. If we knew the magnitude of the force (that is, how many kN) that each of the two men was putting into the exercise, we could calculate the *magnitude* – and exact *direction* – of the resultant force.

Of course, we may have more than two forces. Donald and Tristan's mutual friend Tarquin may enter the room and start pushing the trolley in a different direction while Donald and Tristan are exerting themselves. The trolley would move off in a different direction. Again, the direction of movement of the trolley represents the direction of the resultant force.

Resultants of forces

The resultant force (let's call it R) is the single force that would have the same effect on an object as a system of two or more forces. Resultants can be calculated by simple trigonometry or by graphical methods. Example 7.1 shows how trigonometry may be used.

Example 7.1

An object is subjected to three forces of different magnitudes, acting in different directions, as shown in Fig. 7.4 (a). Using a graphical approach, determine the magnitude and direction of the resultant force.

We can consider the forces in any order. We will consider the vertical 4 kN force first. Using graph paper and choosing a suitable scale (1 cm = 1 kN in this case), we will start from a point A. The vertical force can be represented by a line vertically upwards from point A, 4 cm long (to represent 4 kN). The point we arrive at will be called point B (see Fig. 7.4 (b)). Next, let's consider the horizontal 4 kN force, which acts to the right. Starting from point B, draw a horizontal line 4 cm long (going to the right), representing the horizontal force of 4 kN. The point we arrive at will be point C.

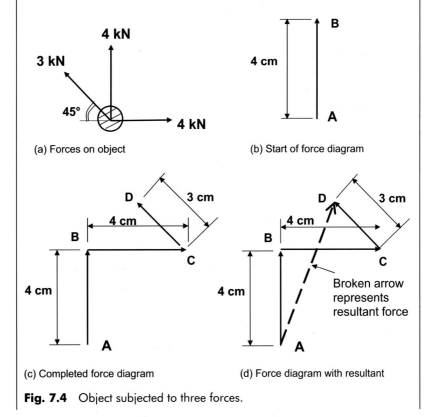

Fig. 7.4 Object subjected to three forces.

Finally, let's consider the 3 kN force, which acts diagonally upwards to the left at an angle of 45 degrees. Starting from point C, this force will be represented by a line 3 cm long (representing 3 kN) in the appropriate direction. The point we arrive at will be point D (see Fig. 7.4 (c)). Next, draw a straight line connecting points A and D. This line represents the resultant force. Measuring off the diagram (Fig. 7.4 (d)) it can be found that the line is 6.41 cm long at an angle to the horizontal of 72.9 degrees.

Therefore the resultant force is 6.41 kN, acting at an angle of 72.9 degrees to the horizontal (upwards and to the right). This is the single force that would have the same effect as the original three forces acting together.

This problem could alternatively have been approached mathematically, using Pythagoras' theorem and basic trigonometry, which are summarised in Appendix 3. The mathematical solution to this problem is shown in Fig. 7.5.

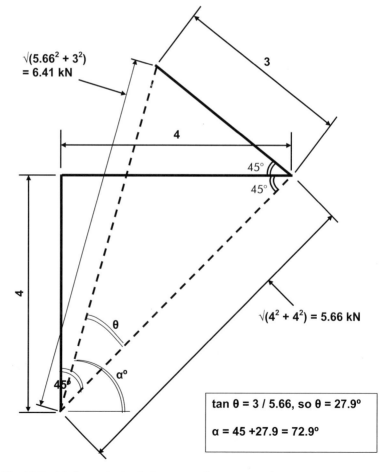

Fig. 7.5 Mathematical solution to resultants example.

Further examples

Each of the examples shown in Fig. 7.6 comprises two forces at right angles to each other. In each case, the task is to find the magnitude and direction of the resultant force. This can be calculated either mathematically or graphically.

To determine the resultants mathematically you will need to reacquaint yourself with the basic mathematics associated with a right-angled triangle, namely Pythagoras' theorem, and the definitions of the trigonometrical functions known as sine, cosine and tangent. Appendix 3 gives you a quick refresher on these. To determine the resultants graphically, you need to represent the forces by lines on graph paper whose lengths are proportional to the magnitudes of the forces. The lines need to be orientated in the same directions as the corresponding forces.

Whichever way you determine the resultant, it doesn't matter which order you consider the forces in; you will still get the same answer – this was the point made by the 'Underground analogy' earlier in the chapter when we saw that there is more than one route from Green Park to Oxford Circus. However, you *must* consider the forces in the 'nose to tail' manner adopted in the example above (and shown in Fig. 7.4 (c)), otherwise your answer for the direction will be wrong. In my experience, by far the most common mistake students make when dealing with this sort of problem lies in their not redrawing the forces 'nose to tail'. So make sure that the 'nose' (that is, the arrowed end) of each force is laid adjacent to the 'tail' (the non-arrowed end) of the next force because 'nose to nose' or 'tail to tail' will give the wrong answer.

Study the worked example given at the top of Fig. 7.6. Note how it has been solved, first of all by expressing the forces in a 'nose to tail' manner, then by calculating the magnitude of the resultant force using Pythagoras' theorem and its direction using trigonometry.

Now attempt the five examples given in Fig. 7.6. Figure 7.7 shows the solutions to the examples shown in Fig. 7.6.

If you got the direction of the first two examples wrong, then you haven't been expressing the forces in the 'nose to tail' manner required; in each case, the problem has to be reconstructed in the manner shown in Fig. 7.7. If you got the first two examples right but came to a dead halt when you reached example number 3, then your mathematical knowledge of right-angled triangles is probably fine but you've lost sight of what resultants actually are. Remember: to obtain the resultant of two or more forces you express the forces (in any order) in a nose to tail fashion, then you draw a line linking the tail of the first force with the nose of the final force. In the case of example 3, this resultant force turns out to be vertically upwards.

You should have realised that examples 4 and 5 can be simplified. For instance, in example 4, the 16 kN force to the right is partially cancelled by the 12 kN force to the left, to give an overall force to the right of 4 kN (i.e. 16 – 12). Similarly, the upward force will be 2 kN (i.e. 10 – 8).

You will find further examples at the end of this chapter.

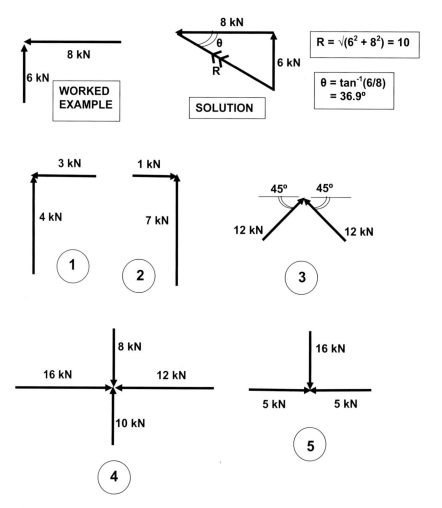

Fig. 7.6 Components: examples.

Components of forces

Earlier in this chapter we looked at how we could express a number of different forces, acting together, at the same point, as a single force – the resultant. Now we are going to invert the process by taking a single force and breaking it down into two forces which, taken together, have the same effect as the original single force.

These two forces are called *components*. In the same way as a television set contains many electrical components, all of which must be present for the television to work, so must both of our force components be present to correctly represent the original force. A component is the replacement of an original force with two forces at right angles to each other (usually one horizontal and one vertical).

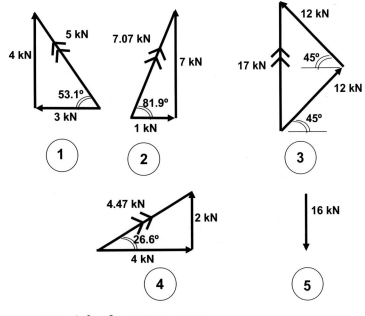

1. $R = \sqrt{(3^2 + 4^2)} = 5$ kN; $\theta = 53.1°$
2. $R = \sqrt{(1^2 + 7^2)} = 7.07$ kN; $\theta = 81.9°$
3. $R = \sqrt{(12^2 + 12^2)} = 17$ kN; $\theta = 90°$ (i.e. vertically upwards)
4. $R = \sqrt{(2^2 + 4^2)} = 4.47$ kN; $\theta = 26.6°$
5. $R = 16$ kN (by inspection), vertically downwards.

Fig. 7.7 Components: solutions to examples.

It can be shown that, for any force F at an angle θ to the horizontal, the horizontal component is always $F.\cos\theta$ and the vertical component is always $F.\sin\theta$ (see Fig. 7.8). As a mnemonic, think of 'sign up' to represent sine being the vertical force.

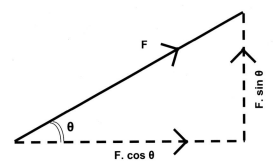

Fig. 7.8 Components of forces: general case.

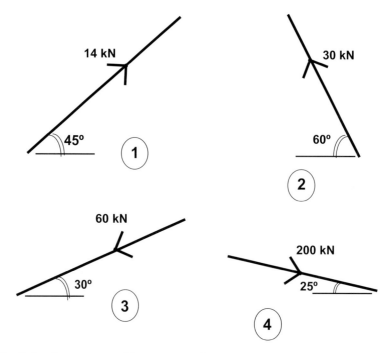

Fig. 7.9 Components of forces: examples.

For each of the three examples given in Fig. 7.9, calculate the magnitude and direction of the two components (one horizontal, the other vertical) of the force given. In each case, make sure you correctly identify whether the horizontal force is to the left or to the right and whether the vertical force is upwards or downwards – such things are important! Check your answers with the following (where H = horizontal component and V = vertical component):

(1) $H = 14.\cos 45° = 9.9$ kN →, $V = 14.\sin 45° = 9.9$ kN ↑.
(2) $H = 30.\cos 60° = 15$ kN ←, $V = 30.\sin 60° = 26$ kN ↑.
(3) $H = 60.\cos 30° = 52$ kN ←, $V = 60.\sin 30° = 30$ kN ↓.
(4) $H = 20.\cos 25° = 181$ kN →, $V = 200.\sin 25° = 84.5$ kN ↓.

We shall see later in this book how useful it is to be able to replace a force acting at an angle by two forces: one horizontal and the other vertical.

What you should remember from this chapter

- The resultant of a number of forces acting at a point is the single force which has the same effect as the original forces acting together.
- Any force can be split into two components which, acting together, have the same effect as the original single force. The two components are at right angles to each other and are usually taken as horizontal and vertical respectively.

More about forces: resultants and components 59

Tutorial examples

Find the resultant for each of the multi-force examples shown in Fig. 7.10 and split each of the one-force examples into components.

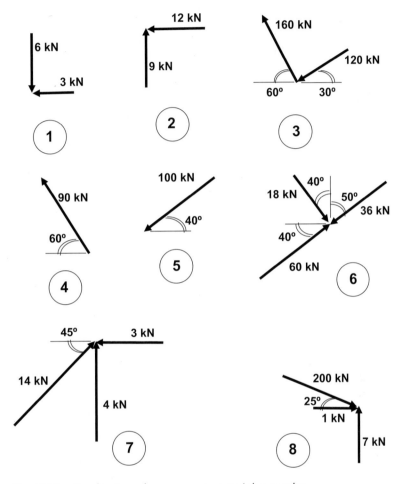

Fig. 7.10 Resultants and components: tutorial examples.

Moments

Introduction

When you use a spanner to tighten or loosen a nut, you are applying a *moment*. A moment is a turning effect and is related to the concept of leverage: if you use a screwdriver to prise open the lid on a can of paint, or a bottle opener to open a bottle of beer, or a crowbar to lift a manhole cover, you are applying leverage and hence you are applying a moment.

Designed by British architect Sir Norman Foster, the modern glass dome shown in Fig. 8.1 is the historic German parliament building's crowning glory.

Fig. 8.1 Reichstag dome, Berlin.

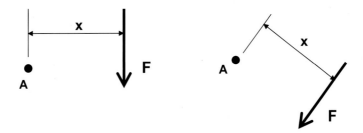

Fig. 8.2 Moments illustrated.

What is a moment?

A moment is a turning effect. A moment always acts about a given point and is either clockwise or anticlockwise in nature. The moment about a point A caused by a particular force F is defined as the force F multiplied by the perpendicular distance from the force's line of action to the point.

Units of moment are kN.m or N.mm Note: This follows because a moment is a force multiplied by a distance, therefore its units are the units of force (kN or N) multiplied by distance (m or mm) – hence kN.m or N.mm. The units of moment are *never* kN or kN/m.

In both the cases illustrated in Fig. 8.2, if M is the moment about point A, then $M = F.x$.

Practical examples of moments

See-saw

A see-saw is a piece of equipment often found in children's playgrounds. It comprises a long plank of wood with a seat at each end. The plank of wood is supported at its centre point. The support is pivoted, so is free to rotate. A child sits on the seat at each end of the see-saw and uses the pivoting characteristic of the see-saw to move up and down. A series of modern see-saws is shown in Fig. 8.3.

Imagine two young children sitting at opposite ends of a see-saw, as shown in Fig. 8.4 (a). If the two children are of equal weight and sitting at equal distances from the see-saw's pivot point, there will be no movement because the clockwise moment about the pivot due to the child at the right-hand end ($F.x$) is equal to the anticlockwise moment about the pivot point due to the child at the left-hand end ($F.x$). Therefore the two moments cancel each other out.

If the child at the left-hand end was replaced by an adult or a much larger child, as shown in Fig. 8.4 (b), the child at the right-hand end would move rapidly upwards. This is because the (anticlockwise) moment due to the larger person at the left-hand end (Large Force × Distance) is greater than the (clockwise) moment due to the small child at the right-hand end (Small Force × Distance). The overall moment is thus anticlockwise, causing upward movement of the small child.

Fig. 8.3 See-saws.

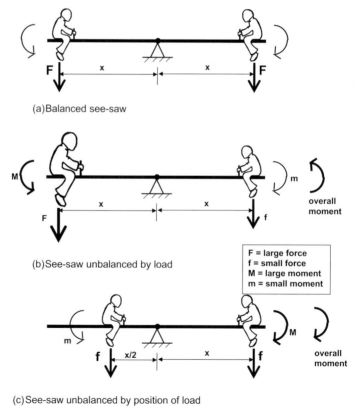

(a) Balanced see-saw

(b) See-saw unbalanced by load

F = large force
f = small force
M = large moment
m = small moment

(c) See-saw unbalanced by position of load

Fig. 8.4 Forces on a see-saw.

Let's return to the original situation, with two young children at opposite ends of the see-saw. But suppose now that the left-hand child moved closer to the pivot point, as shown in Fig.8.4 (c). As a result the right-hand child would move downwards. This is because the anticlockwise moment due to the left-hand child (Force × Small Distance) is smaller than the clockwise moment due to the right-hand child (Force × Large Distance). The overall movement is thus clockwise, causing downward movement of the right-hand child.

Spanners, nuts and bolts

The reader will know from experience that it is much easier to undo a seized-up nut or bolt if a long spanner is used rather than a short spanner. This is because, although the force used may be the same, the 'lever arm' distance is longer, thus causing a greater turning effect or moment to be applied. Practical problems using 'leverage' also illustrate this principle, such as the examples of prising open paint cans, beer bottles and manhole covers already mentioned.

Numerical problems involving moments

It can be seen from the above that it is important to distinguish between clockwise and anticlockwise moments. After all, turning a spanner clockwise (tightening a nut) has a very different effect from turning the spanner anticlockwise (loosening a nut). In this book:

- clockwise moments are regarded as positive (+)
- anticlockwise moments are regarded as negative (−).

It is, of course, quite possible for a given pivot point to experience several moments simultaneously, some of which may be clockwise (+) while others may be anticlockwise (−). In these cases, moments must be added algebraically to obtain a total (net) moment.

Some simple worked examples of moment calculation

In each of the following examples, involving simple beams, we are going to calculate the net moment about point A (remember, clockwise is +, anticlockwise is −).

Example 1 (see Fig. 8.5 (a))
By inspection, the 4 kN force is trying to turn clockwise about A, therefore the moment will be positive (+). The 2 metre distance is measured horizontally from the (vertical) line of action of the 4 kN force; in other words, the distance given is measured perpendicular (i.e. at right angles or 90 degrees) to the line of action of the force, as required.

Remember, a moment is a force multiplied by a distance. If we use the symbol M to represent moment, then in this case:

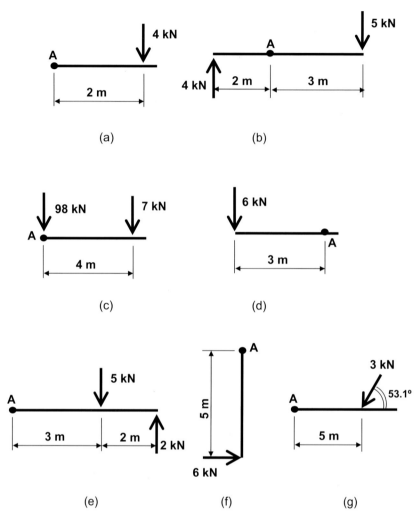

Fig. 8.5 Moment worked examples.

$M = + (4 \text{ kN} \times 2 \text{ m}) = + 8 \text{ kN.m}$

Example 2 (see Fig. 8.5 (b))
This time there are two forces, supplying two moments. A common mistake with this example is to assume that since the two forces are in opposite directions (i.e. one upwards, one downwards), the moments must also oppose each other. In fact, a closer inspection will reveal that the moments about A generated by the two forces are both clockwise (+). So the moment about A for each force is calculated, and the two added together, as follows:

$M = + (5 \text{ kN} \times 3 \text{ m}) + (4 \text{ kN} \times 2 \text{ m})$

$$= +15 \text{ kN.m} + 8 \text{ kN.m}$$
$$= +23 \text{ kN.m}$$

(If you attempted this example and obtained an answer of + 7 kN.m, you fell into the trap mentioned above!)

Example 3 (see Fig. 8.5 (c))
Once again there are two forces, supplying two moments. The 7 kN force clearly gives rise to a clockwise moment about A. The 98 kN force, however, passes straight through the pivot point A; in other words, its line of action is zero distance from A. Since a moment is always a force multiplied by a distance, if the distance is zero then it follows that the moment must be zero (since multiplying any number by zero gives a product of zero). So, in this example:

$$M = +(7 \text{ kN} \times 4 \text{ m}) + (98 \text{ kN} \times 0 \text{ m})$$
$$= +28 \text{ kN.m} + 0 \text{ kN.m}$$
$$= +28 \text{ kN.m}$$

The lesson to be learned from this example is: if a force passes through a certain point, then the moment of that force about that point is zero.

Example 4 (see Fig. 8.5 (d))
The 6 kN force is turning anticlockwise about A, so the resulting moment will be negative (−).

$$M = -(6 \text{ kN} \times 3 \text{ m}) = -18 \text{ kN.m}$$

Example 5 (see Fig. 8.5 (e))
The 5 kN force is trying to turn clockwise about A, therefore will give rise to a clockwise (+) moment. By contrast, the 2 kN force is trying to turn anticlockwise about A, therefore will produce an anticlockwise (−) moment.

$$M = +(5 \text{ kN} \times 3 \text{ m}) - (2 \text{ kN} \times 5 \text{ m})$$
$$= +15 \text{ kN.m} - 10 \text{ kN.m}$$
$$= +5 \text{ kN.m}$$

Example 6 (see Fig. 8.5 (f))
Not all forces are vertical! But the same rules apply.

$$M = -(6 \text{ kN} \times 5 \text{ m}) = -30 \text{ kN.m}$$

Example 7 (see Fig. 8.5 (g))
This slightly harder example will confuse readers who haven't yet grasped the fact that a moment is a force multiplied by a *perpendicular* (or 'lever arm') distance. There are two ways of solving this problem – see Fig. 8.6.

1. By using trigonometry to find the perpendicular distance. Figure 8.6 (a) will remind you of the definitions of sines, cosines and tangents in terms of the lengths of the sides of a right-angled triangle. Applying

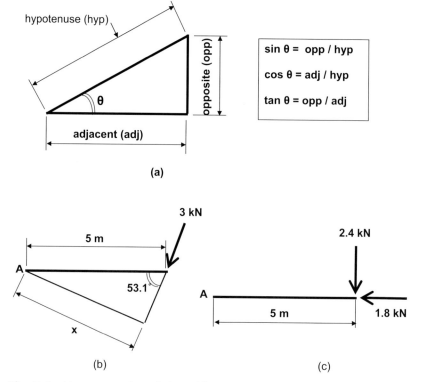

Fig. 8.6 Moments and resolution of forces.

this to the current problem, we find from Fig. 8.6 (b) that the perpendicular distance (x) in this case is 4 metres. So $M = + (3 \text{ kN} \times 4 \text{ m}) = + 12$ kN.m.

2 By resolving the 3 kN force into vertical and horizontal components. In Chapter 7 we learned that any force can be expressed as the product of two components, one horizontal and one vertical. For any force F acting at an angle of θ to the horizontal axis, it can be shown that:
- the horizontal component is always $F \times \cos \theta$, and
- the vertical component is always $F \times \sin \theta$ (so sin acts upwards: 'Sign Up').

In this problem the 3 kN force acts at an angle of 53.1 degrees to the horizontal. So its vertical component = $3 \times \sin 53.1° = 2.4$ kN ↓
And its horizontal component = $3 \times \cos 53.1° = 1.8$ kN. ←
The problem can now be expressed as shown in Fig. 8.6 (c).

Note that since the 1.8 kN force (extended) passes through point A, the moment of that force about point A will be zero.

$M = + (2.4 \text{ kN} \times 5 \text{ m}) + (1.8 \text{ kN} \times 0)$
$= + 12$ kN.m

(Obviously, this is the same answer as that obtained in part (a)!)

There are some further examples at the end of the chapter for you to try.

Notes on moment calculations

From the above discussion and examples come the following notes and observations:

(1) Always consider whether a given moment is clockwise (+) or anti-clockwise (−).
(2) If a force F passes through a point A, then the moment of force F about point A is zero (as illustrated in Example 3 above).
(3) It may be necessary to resolve forces into components in order to calculate moments (as shown in Example 6 above).

Moment equilibrium

Imagine that you are doing some mechanical work on a car and you have a spanner fitted onto a particular bolt in the car's engine compartment. You are trying to tighten the nut and hence you are turning it clockwise. In other words, you are applying a clockwise moment. Now imagine that a friend (in the loosest possible sense of the word!) has another spanner fitted to the same nut and is turning it anticlockwise. This has the effect of loosening the nut.

If the anticlockwise moment that your friend is applying is the same as the clockwise moment that you are applying (regardless of the fact that the two spanners might be of different lengths), you can imagine that the two effects would cancel each other out – in other words, the nut would not move. This applies to any object subjected to equal turning moments in opposite directions: the object would not move.

So if the total clockwise moment about a point equals the total anticlockwise moment about the point, no movement can take place. Conversely, if there is no movement (as is usually the case with a building or any part of a building), then clockwise and anticlockwise moments must be balanced. This is the principle of moment equilibrium and can be used in conjunction with the rules of force equilibrium (discussed earlier) to solve structural problems.

To summarise: if any object (such as a building or any point within a building) is stationary, the net moment at the point will be zero. In other words, clockwise moments about the point will be cancelled out by equal and opposite anticlockwise moments.

Equilibrium revisited

As discussed in Chapter 6, if an object, or a point within a structure, is stationary, we know that forces must balance, as follows:

$\Sigma V = 0$, i.e. Total Upward Force = Total Downward Force (↑ = ↓)

$\Sigma H = 0$, i.e. Total Force to Left = Total Force to Right ($\leftarrow = \rightarrow$)

From our newly acquired knowledge of moments, we can add a third rule of equilibrium, as follows:

$\Sigma M = 0$, i.e. Total Clockwise Moment = Total Anticlockwise Moment

We can use these three rules of equilibrium to solve structural problems, specifically the calculation of end reactions, as discussed in the next chapter.

What you should remember from this chapter

- A moment is one of the most important concepts in structural mechanics.
- A moment is a turning effect, either clockwise or anticlockwise, about a given point.
- If a force passes through the point about which moments are being taken, then the moment of that force about the point concerned is zero. (It is very important to remember this concept as it crops up several times in the solution of problems later in this book.)
- It may be necessary to resolve forces into components in order to calculate moments.

Tutorial examples

In each of the examples shown in Fig. 8.7, calculate the net moment, in kN.m units, about point A.

Tutorial answers

(1) $M = +90$ kN.m
(2) $M = -40$ kN.m
(3) $M = +50$ kN.m
(4) $M = +90$ kN.m
(5) $M = +1$ kN.m
(6) $M = +63$ kN.m

Fig. 8.7 Moment tutorial examples.

Reactions

Note: the support symbols used in the diagrams in this chapter will be explained in Chapter 10.

Introduction

In Chapter 6 we discussed equilibrium. We found out that if a body or object of any sort is stationary, then the forces on it balance, as follows:

Total force upwards = Total force downwards

Total force to the left = Total force to the right

This was illustrated in Fig. 6.4.

The concept of a moment, or turning effect, was introduced in Chapter 2 and discussed more fully in Chapter 8. In this chapter we will find out how to use this information to calculate *reactions* – that is, the upward forces that occur at beam supports in response to the forces on the beam.

Moment equilibrium

At the end of Chapter 8 we found that if an object or body is stationary, it doesn't rotate and the total clockwise moment about any point on the object is equal to the total anticlockwise moment about the same point. This is the third rule of equilibrium and we can add this to the first two that we discovered in Chapter 6. The three rules of equilibrium are expressed in Fig. 9.1.

Figure 9.2 shows a steel-framed building under construction. Note the steel beams, the steel columns and the profiled steel deck flooring.

Reactions   71

Total force up =
total force down
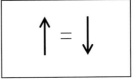

Total force to left =
total force to right

Total clockwise moment =
total anticlockwise moment

Fig. 9.1 The rules of equilibrium.

Fig. 9.2 Steel-framed building under construction.

Calculation of reactions

The three rules of equilibrium can be used to calculate reactions. As discussed in Chapter 2 and again in Chapter 6, a reaction is a force (usually upwards) that occurs at a support of a beam or similar structural element. A reaction counteracts the (usually downward) forces in the structure to maintain equilibrium. It is important to be able to calculate these reactions. If the support is a column, for example, the reaction represents the force in the column, which we would need to know in order to design the column.

Consider the example shown in Fig. 9.3. The thick horizontal line represents a beam of span 6 metres which is simply supported at its two ends, A and B. The only load on the beam is a point load of 18 kN, which acts vertically downwards at a position 4 metres from point A. We are going to calculate the reactions R_A and R_B (that is, the support reactions at points A and B respectively).

From *vertical equilibrium*, which we discussed in Chapter 6, we know that:

Total force upwards = Total force downwards

Applying this to the example shown in Fig. 9.3, we can see that:

$R_A + R_B = 18$ kN

Of course, this doesn't tell us the value of R_A and it doesn't tell us the value of R_B. It merely tells us that the sum of R_A and R_B is 18 kN. To evaluate R_A and R_B then, we clearly have to do something different.

Let's use our new-found knowledge of moment equilibrium. We found out above that if any structure is stationary, then at any given point in the structure:

Total clockwise moment = Total anticlockwise moment

The above applies at any point in a structure. So, taking moments about point A:

$(18 \text{ kN} \times 4 \text{ m}) = (R_B \times 6 \text{ m})$

Fig. 9.3 Calculation of reactions for point loads.

Therefore $R_B = 12$ kN. Note that there is no moment due to force R_A. This is because force R_A passes straight through the point (A) about which we are taking moments.

Similarly, taking moments about point B:

Total clockwise moment = Total anticlockwise moment

$$(R_A \times 6 \text{ m}) = (18 \text{ kN} \times 2 \text{ m})$$

Therefore $R_A = 6$ kN.

As a check, let's add R_A and R_B together:

$$R_A + R_B = 6 + 12 = 18 \text{ kN}$$

which is what we would expect from the first equation above.

A word of warning ...

It's easy to make a mistake and get the two reactions the wrong way round. As a check, consider the man standing on a scaffold board supported by scaffold poles at each end, as shown in Fig. 9.4. The man is standing closer to the left-hand support. Which of the two supports is doing the more work in supporting the man's weight?

Common sense tells us that the left-hand support must be working harder to bear the man's weight, simply because the man is closer to that support. In other words, we would expect the left-hand support reaction to be the greater of the two.

Looking again at the example shown in Fig. 9.3, the 18 kN loading occurs towards the right-hand end of the beam, so we would expect the right-hand end reaction (R_B) to be greater than the left-hand end reaction (R_A). And indeed it is.

Fig. 9.4 Man on a scaffold board.

It's always a good idea to do this 'common sense check' to ensure you've got the reactions the right way round. To summarise: if the loading on the beam is clearly greater at one end of the beam, you would expect the reaction to be greater at that end too.

Calculation of reactions when uniformly distributed loads (UDLs) are present

Up till now in this chapter we have looked only at problems with point loads and have studiously avoided those with uniformly distributed loads. There is a good reason for this: analysis of problems with point loads only is much easier and it has been my policy in writing this book – as in life in general – to start with the easy things and work up to the harder ones.

In practice, most loads in 'real' buildings and other structures are uniformly distributed loads – or can be represented as such – so we need to know how to calculate end reactions for such cases. The main problem we encounter is in taking moments. For point loads it is straightforward – the appropriate moment is calculated by multiplying the load (in kN) by the distance from it to the point about which we're taking moments. However, with a uniformly distributed load, how do we establish the appropriate distance?

Figure 9.5 represents a portion of uniformly distributed load of length x. The intensity of the uniformly distributed load is w kN/m. The chain-dotted line in Fig. 9.5 represents the centre line of the uniformly distributed load. Let's suppose we want to calculate the moment of this piece of UDL about a point A, which is located a distance a from the centre line of the UDL. In this situation, the moment of the UDL about A is the total load multiplied by the distance from the centre line of the UDL to the point about which we're taking moments. The total UDL is $w \times x$, the distance concerned is x, so:

Fig. 9.5 Bending moment calculation for uniformly distributed load (UDL): general case.

moment of UDL about A = wax.

Apply this principle whenever you're working with uniformly distributed loads.

Example involving uniformly distributed loads (UDLs)

Calculate the end reactions for the beam shown in Fig. 9.6. Use the same procedure as before.

Vertical equilibrium:

$$R_A + R_B = (3 \text{ kN/m} \times 2 \text{ m}) = 6 \text{ kN}$$

Taking moments about A:

$$(3 \text{ kN/m} \times 2 \text{ m}) \times 1 \text{ m} = R_B \times 4 \text{ m}$$

Therefore:

$$R_B = 1.5 \text{ kN}$$

Taking moments about B:

$$(3 \text{ kN/m} \times 2 \text{ m}) \times 3 \text{ m} = R_A \times 4 \text{ m}$$

Therefore:

$$R_A = 4.5 \text{ kN}$$

Check:

$$R_A + R_B = 4.5 + 1.5 = 6 \text{ kN}$$

(as expected from the first equation).

What you should remember from this chapter

The third rule of equilibrium tells us that if an object, or any part of it, is stationary, the total clockwise moment about any point within the object is the same as the total anticlockwise moment about that point. This rule can

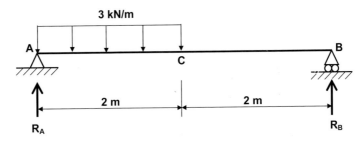

Fig. 9.6 Calculation of reactions for uniformly distributed loads.

76 Basic Structures for Engineers and Architects

be used in conjunction with the first two rules of equilibrium to calculate support reactions.

Tutorial examples

Try the examples given in Fig. 9.7. In each case, calculate the reactions at the support positions.

Tutorial answers

(a) $R_A = 75$ kN, $R_B = 45$ kN
(b) $R_A = 7.5$ kN, $R_B = 16.5$ kN
(c) $R_A = 17.5$ kN, $R_B = 22.5$ kN
(d) $R_A = 17.5$ kN, $R_B = 12.5$ kN

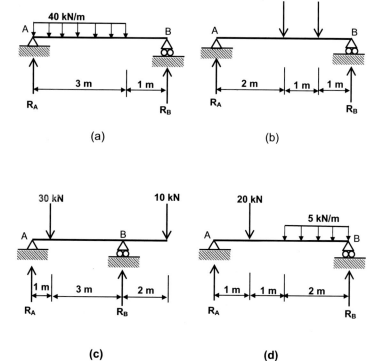

Fig. 9.7 Reactions: further tutorial examples.

10 Different types of support – and what's a pin?

What is a PIN?

I would be flattered to think that you are reading this book on a beach, at a rural beauty spot or in some other glamorous location. But the chances are you are inside a building as you skim through these words – perhaps your home, office or college – in which case you will probably be in sight of a door. If it's a conventional door (not a sliding one, for example), it will have hinges on it. What are the hinges for? Well, they make it possible for you to open the door by rotating it about the vertical axis on which the hinges are located.

Figures 10.1 (a) and 10.1 (b) show the plan view of a door, in its shut and partially open positions respectively, along with part of the adjoining wall. You could approach this door and open it or shut it, partially or totally, at will. The hinges make it possible for you to do this by facilitating rotation. Had the door been rigidly fixed to the wall you would not have been able to open it at all. One other point to note: although you can open or shut the door at will, nothing you do to the door will affect the portion of the wall on the other side of the hinges. It remains unmoved. To put it another way, the hinges do not transmit rotational movements into the wall. This is a particularly important concept and is the basis of the analysis of pin-jointed frames, which we will investigate in Chapters 12–15.

(a) Closed door   (b) Open door

Fig. 10.1 A door viewed from above.

(a) two rods in line (b) left rod rotated

Fig. 10.2 Steel rods connected by a pin.

The word *pin*, as used in structural engineering, is analogous to the hinge in a door. A pin is indicated symbolically as a small unfilled circle. Consider two steel rods connected by a pin joint, as shown in Fig. 10.2. The two rods are initially in line as shown in Fig. 10.2 (a) and the left-hand rod is subsequently rotated about 30 degrees anticlockwise, as shown in Fig. 10.2 (b). The right-hand rod is not affected by this rotational movement of the left-hand rod.

A pin, then, has two important characteristics:

(1) A pin permits *rotational* movement about itself.
(2) A pin cannot *transmit* turning effects, or moments.

Different types of support

Up till now we've been talking about supports (to beams, etc.) and indicating them as upward arrows without giving any thought to the type or nature of the support. As we shall see, there are three different types of support.

1. Roller supports

Imagine a person on roller skates standing in the middle of a highly polished floor. If you were to approach this person and give him (or her) a sharp push from behind (not to be recommended without discussing it with them first!), they would move off in the direction you pushed them. Because they are on roller skates on a smooth floor, there would be minimal friction to resist the person's slide across the floor.

A *roller support* to part of a structure is analogous to that person on roller skates: a roller support is free to move horizontally. Roller supports are indicated using the symbol shown in Fig. 10.3 (a). You should recognise that this is purely symbolic and a real roller support will probably not resemble this symbol. In practice a roller support might comprise sliding rubber bearings, for example, or steel rollers sandwiched between steel plates, as shown in Fig. 10.3 (b).

2. Pinned supports

Consider the door hinge analogy discussed above. A *pinned support* permits rotation but cannot move horizontally or vertically – in exactly the

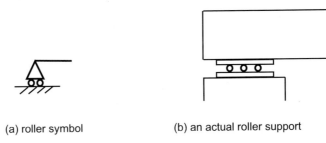

(a) roller symbol   (b) an actual roller support

Fig. 10.3 Roller support – symbolically and in reality.

same way as a door hinge provides rotation but cannot itself move from its position in any direction.

3. Fixed supports

Form your two hands into fists, place them about a foot apart horizontally and allow a friend to position a ruler on your two fists so that it is spanning between them. Your fists are safely supporting the ruler at each end. Now remove one of the supports by moving your fist out from underneath the ruler. What happens? The ruler drops to the floor. Why? You have removed one of the supports and the remaining single support is not capable of supporting the ruler on its own – see Figs 10.4 (a) and (b).

However, if you grip the ruler between your thumb and remaining fingers at one end only, it can be held horizontally without collapsing. This is because the firm grip provided by your hand prevents the end of the ruler from rotating and thus falling to the floor – see Fig. 10.4 (c).

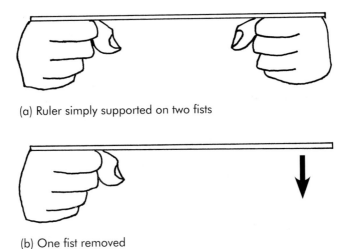

(a) Ruler simply supported on two fists

(b) One fist removed

Fig. 10.4 What is a fixed support? (*Continued.*)

(c) Ruler firmly gripped at one end

Fig. 10.4 (Continued.)

In structures, the support equivalent to your gripping hand in the above example is called a *fixed support*. As with your hand gripping the ruler, a fixed support does not permit rotation.

There are many situations in practice where it is necessary (or at least desirable) for a beam or slab to be supported at one end only – for example, a balcony. In these situations, the single end support must be a fixed support because, as we've seen, a fixed support does not permit rotation and hence does not lead to collapse of the structural member concerned – see Fig. 10.5. Like a pinned support, a fixed support cannot move in any direction from its position. Unlike a pinned support, a fixed support cannot rotate. So a fixed support is fixed in every respect.

Now you've got a mental picture of each of the three different types of support (roller, pinned and fixed), let's revisit each of them and take our study of them a stage further. We are going to do this in the context of reactions and moments.

Restraints

Let's consider each of the following as being a restraint:

(1) Vertical reaction
(2) Horizontal reaction
(3) Resisting moment.

(a) fixed support symbol   (b) an actual fixed support

Fig. 10.5 Fixed support – symbolically and in reality.

Restraints experienced by different types of support

Roller support
Let's return to our roller skater standing on a highly polished floor. As the floor is supporting him, it must be providing an upward reaction to counteract the weight of the skater's body. However, we've already seen that if we push our skater, he will move. The rollers on the skates, and the frictionless nature of the floor, mean that the skater can offer no resistance to our push. In other words, the skater can provide no horizontal reaction to our pushing (in contrast to a solid wall, for example, which would not move if leaned on and therefore would provide a horizontal reaction). There is also nothing to stop the skater from falling over (i.e. rotating).

We can see from the above that a roller support provides one restraint only: *vertical reaction*. (There is no horizontal reaction and no moment.)

Pinned support
As discussed above, a pinned support permits rotation (so there is no resistance to moment), but as it cannot move horizontally or vertically there must be both horizontal and vertical reactions present. So, a pinned support provides two restraints: *vertical reaction* and *horizontal reaction*. (There is no moment.)

Fixed support
We saw above that a fixed support is fixed in every respect: it cannot move either horizontally or vertically and it cannot rotate. This means there will be both horizontal and vertical reactions and, if it cannot rotate, there must be a moment associated with the fixed support. Incidentally, this moment is called a *fixed end moment* – see Chapter 8 if you are not clear on this point.

So, a fixed support provides three restraints: vertical reaction, horizontal reaction and moment.

To summarise:

- A roller support provides one restraint: vertical reaction.
- A pinned support provides two restraints: vertical reaction and horizontal reaction.
- A fixed support provides three restraints: vertical reaction, horizontal reaction and moment.

This is illustrated in Fig. 10.6.

Simultaneous equations

Let's brush up our knowledge of mathematics for a few minutes – specifically, equations and simultaneous equations.

Answer the following question with a simple Yes or No: can you solve the following equation?

$x + 6 = 14$

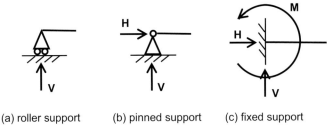

(a) roller support   (b) pinned support   (c) fixed support

Fig. 10.6 Restraints provided by various support types.

Clearly, the answer is Yes. You can solve the above equation very easily ($x = 8$), but why? The reason you can solve the above equation so easily is that there is only one unknown (x in this case).

Now consider whether you can solve the following two simultaneous equations:

$$2x + 6y = -22$$

$$3x - 4y = 19$$

Again, it is possible to solve these two equations (although you may need to brush up your maths in order to do so!). The solution, incidentally, is $x = 1, y = -4$. Again, why is it possible to solve these equations? The reason this time is that, although we have two unknowns (x and y), we have two equations.

Now consider whether or not you could solve the following simultaneous equations:

$$4x + 2y - 3z = 78$$

$$2x - y + z = 34$$

If you haven't realised for yourself, I'll spare you the tedium of trying to work it out by telling you that No, you can't solve the problem in this case. The reason is that this time we have three unknowns (x, y and z), but only two equations.

We could carry on investigating in this vein for some time and if we were to do so we would find out the following:

- If we have the same number of unknowns as we have equations, a mathematical problem can be solved.
- But if we have more unknowns than equations, a mathematical problem cannot be solved.

Relating this to structural analysis, if we look back to the procedure we used for calculating reactions in Chapter 9, we'll see that we were solving three equations. These equations were represented by:

(1) Vertical equilibrium (total force up = total force down)
(2) Horizontal equilibrium (total force right = total force left)
(3) Moment equilibrium (total clockwise moment = total anticlockwise moment).

Different types of support – and what's a pin?   83

As we have three equations, we can use them to solve a problem with up to three unknowns in it. In this context, an unknown is represented by a restraint, as defined earlier in this chapter. (Remember, a roller support has one restraint, a pinned support has two restraints and a fixed support has three restraints.) So a structural system with up to three restraints is solvable – such a system is said to be statically determinate (SD) – while a structural system with more than three restraints is not solvable (unless we use advanced structural techniques which are well beyond the scope of this book) – such a system is said to be statically indeterminate (SI).

So if we inspect a simple structure, examine its support and thence count up the number of restraints, we can determine whether the structure is statically determinate (up to three restraints in total) or statically indeterminate (more than three restraints).

Let's look at the three examples shown in Fig. 10.7.

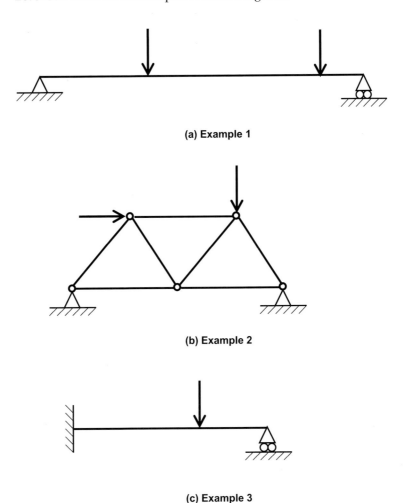

(a) Example 1

(b) Example 2

(c) Example 3

Fig. 10.7   Statical determinacy.

Example 1

> This beam has a pinned support (two restraints) at its left-hand end and a roller support (one restraint) at its right-hand end. So the total number of restraints is (2 + 1) = 3, therefore the problem is solvable and is statically determinate (SD).

Example 2

> This pin-jointed frame has a pinned support (two restraints) at each end. So the total number of restraints is (2 + 2) = 4. As 4 is greater than 3, the problem is not solvable and is statically indeterminate (SI).

Example 3

> This beam has a fixed support (three restraints) at its left-hand end and a roller support (one restraint) at its right-hand end. So the total number of restraints is (3 + 1) = 4, therefore, again, the problem is not solvable and is statically indeterminate (SI).

What you should remember from this chapter

- Supports to structures are one of three types: roller, pinned or fixed. Each provides a certain degree of restraint to the structure at that point.
- Knowing the number of supports a structure has and the nature of each support, it can be determined whether the structure is statically determinate (SD) or statically indeterminate (SI).
- A statically determinate (SD) structure is one that can be analysed using the principles of equilibrium discussed in the earlier chapters of this book. A statically indeterminate (SI) structure cannot be analysed using such principles.

Figure 10.8 shows an 'interim' railway station which was opened in 2004 to allow for repairs to the original St Pancras station in London. Note that the roof and its support structure are completely independent of the vertical glazing structure.

Tutorial examples

> Determine whether each of the structures given in Fig. 10.9 is statically determinate (SD) or statically indeterminate (SI).

Different types of support – and what's a pin? **85**

Fig. 10.8 St Pancras 'interim' railway station, London.

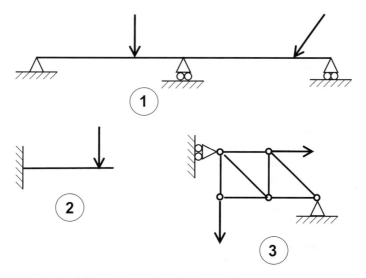

Fig. 10.9 Statical determinacy – tutorial examples.

A few words about stability

Introduction

It is essential for a structure to be strong enough to be able to carry the loads and moments to which it will be subjected. But strength is not sufficient: the structure must also be stable.

In this chapter we'll be looking at what constitutes stability in structural terms – and how we can determine whether or not a particular structural framework is stable. Then we'll look, in practical terms, at how stability is achieved and ensured in buildings.

Stability of structural frameworks

Many buildings and other structures have a structural frame. Steel buildings comprise a framework, or skeleton, of steel. If you live in or near a large city you will have seen such frameworks being constructed. Many bridges also have a steel framework – famous examples include the Tyne Bridge in Newcastle upon Tyne and the Sydney Harbour Bridge in Australia.

We are going to consider the build-up of a framework from scratch. Our framework will consist of metal rods ('members') joined together at their ends by pins. (The concept of a pin, which is a type of connection that facilitates rotation, was discussed in Chapter 10.) Consider two members connected by a pin joint, as shown in Fig. 11.1 (a). Is this a stable structure? (In other words, is it possible for the two members to move relative to each other?) As the pin allows the two members to move relative to one another, this is clearly not a stable structure.

Now, let's add a third member to obtain three members connected by pin joints to form a triangle, as shown in Fig. 11.1 (b). Is this a stable structure? Yes, it is because even though the joints are pinned, movement of the three members relative to each other is not possible. So this is a stable, rigid

A few words about stability   87

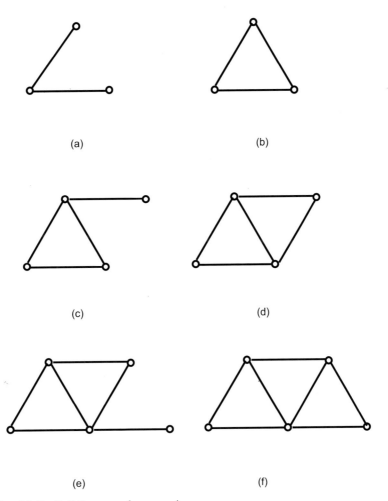

Fig. 11.1 Building up a framework.

structure. In fact, the triangle is the most basic stable structure, as we will mention again in the following discussion.

If we add a fourth member we produce the frame shown in Fig. 11.1 (c). Is this a stable structure? No it is not. Even though the triangle within it is stable, the 'spur' member is free to rotate relative to the triangle, so overall this is not a stable structure.

Consider the frame shown in Fig. 11.1 (d), which is achieved by adding a fifth member to the previous frame. This is a stable structure. If you are unsure of this, try to determine which individual member(s) within the frame can move relative to the rest of the frame. You should see that none of them can and therefore this is a stable structure. This is why you often see this detail in structural frames as 'diagonal bracing', which helps to ensure the overall stability of a structure.

Table 11.1 Is a structure stable?

	m	j	Stable structure?	$2j - 3$	Is $m = 2j - 3$?
Figure 11.1 (a)	2	3	No	3	No
Figure 11.1 (b)	3	3	Yes	3	Yes
Figure 11.1 (c)	4	4	No	5	No
Figure 11.1 (d)	5	4	Yes	5	Yes
Figure 11.1 (e)	6	5	No	7	No
Figure 11.1 (f)	7	5	Yes	7	Yes

Let's add yet another member to obtain the frame shown in Fig. 11.1 (e). Is this a stable structure? No, it is not. In a similar manner to the frame depicted in Fig. 11.1 (c), it has a spur member which is free to rotate relative to the rest of the structure. Adding a further member we can obtain the frame shown in Fig. 11.1 (f) and we will see that this is a rigid, or stable, structure.

We could carry on ad infinitum in this vein, but I think you can see that a certain pattern is emerging. The most basic stable structure is a triangle (Fig. 11.1 (b)). We can add two members to a triangle to obtain a 'new' triangle. All of the frames that comprise a series of triangles (Figs 11.1 (d) and (f)) are stable; the remaining ones, which have spur members, are not.

Let's now see whether we can devise a means of predicting mathematically whether a given frame is stable or not. In Table 11.1 each of the six frames considered in Fig. 11.1 is assessed. The letter m represents the number of members in the frame and j represents the number of joints (note that unconnected free ends of members are also considered as joints). The column headed 'Stable structure?' merely records whether the frame is stable ('Yes') or not ('No').

It can be shown that if $m = 2j - 3$ then the structure is stable. If that equation does not hold, then the structure is not stable. This is borne out by Table 11.1: compare the entries in the column headed 'Stable structure?' with those in the column headed 'Is $m = 2j - 3$?'.

Internal stability of framed structures – a summary

(1) A framework which contains exactly the correct number of members required to keep it stable is termed a ***perfect frame***. In these cases, $m = 2j - 3$, where m is the number of members in the frame and j is the number of joints (including free ends). Frames (b), (d) and (f) in Fig. 11.1 are examples.

(2) A framework having less than the required number of members is unstable and is termed a ***mechanism***. In these cases, $m < 2j - 3$. Frames (a), (c) and (e) in Fig. 11.1 are examples. In each case, one member of the frame is free to move relative to the others.

(3) A framework having more than this required number is 'over-stable' and contains redundant members that could (in theory at least) be removed. Examples follow. In these cases, $m > 2j - 3$. These frames are

statically indeterminate (SI). We met this term in Chapter 10 – it means that the frames cannot be mathematically analysed without resorting to advanced structural techniques.

Examples

For each of the frames shown in Fig. 11.2, use the equation $m = 2j - 3$ to determine whether the frame is (a) a perfect frame (SD), (b) a mechanism (Mech) or (c) statically indeterminate (SI). Where the frame is a mechanism, indicate the manner in which the frame could deform. Where the frame is statically indeterminate, consider which members could be re-

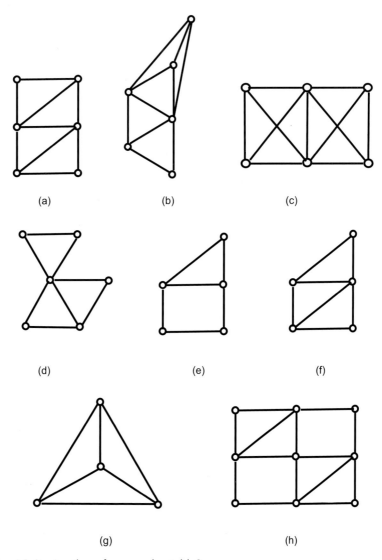

Fig. 11.2 Are these frameworks stable?

Table 11.2 Stability of frames shown in Fig. 11.2

	m	j	2j – 3	Is m = 2j – 3? (or > or <)	Stability type
Figure 11.2 (a)	9	6	9	=	SD
Figure 11.2 (b)	10	6	9	>	SI
Figure 11.2 (c)	11	6	9	>	SI
Figure 11.2 (d)	8	6	9	<	Mech
Figure 11.2 (e)	6	5	7	<	Mech
Figure 11.2 (f)	7	5	7	=	SD
Figure 11.2 (g)	6	4	5	>	SI
Figure 11.2 (h)	14	9	15	<	Mech

moved without affecting the stability of the structure. The answers are given in Table 11.2.

The frames shown in Figs 11.2 (b), (c) and (g) are statically indeterminate. This means they are over-stable and that one or more members may be removed without compromising stability. In the case of Fig. 11.2 (b), any one member can be removed from the top part of the frame and the structure would still be stable. In Fig. 11.2 (c), two members could be removed without compromising stability – but the two members to be removed should be chosen with care. A sensible choice would be to remove one diagonal member from each of the two squares. In Fig. 11.2 (g), any one member could be removed.

The frames shown in Figs 11.2 (d), (e) and (h) are mechanisms. This means that a part of the frame is able to move relative to another part of the frame. In Fig. 11.2 (d), the upper triangle is free to rotate about the frame's central pin independently of the lower part of the frame. In Fig. 11.2 (e), the square part of the frame is free to deform, or collapse, as we shall see in a later example.

The mode of deformation of the frame in Fig. 11.2(h) is less easy to visualise. It is shown in Fig. 11.3.

General cases

Look at frames (a) and (b) in Fig. 11.4. If we apply the $m = 2j - 3$ formula to the standard square depicted in Fig. 11.4 (a), we will find that it is unstable, or a mechanism. It can deform in the manner indicated by the broken lines in Fig. 11.4 (a). This is why, in 'real' structures, diagonal cross-bracing must often be provided to ensure stability.

If we look at the frame shown in Fig. 11.4 (b), we see that it is a square which is diagonally cross-braced twice. Applying the $m = 2j - 3$ formula we find that it is statically indeterminate, which means that it contains at least one redundant member. On further investigation we find that we can remove any one of the six members without affecting the stability of the structure.

A few words about stability 91

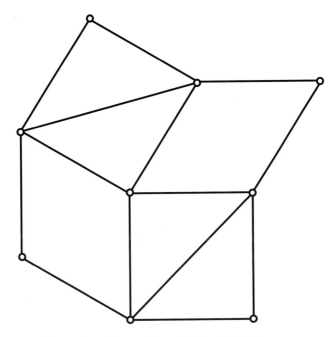

Fig. 11.3 Deformation of frame shown in Fig. 11.2 (h).

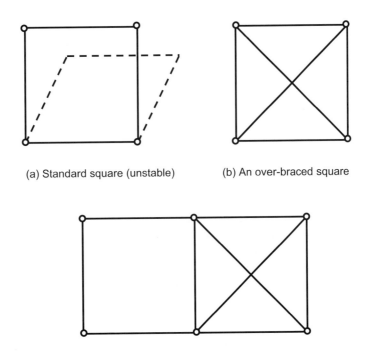

(a) Standard square (unstable)

(b) An over-braced square

(c) Perfect frame, mechanism or over-braced?

Fig. 11.4 Frame stability – general cases.

During the years I've been teaching this subject I've discovered that certain students derive a great deal of comfort from being taught a set of rules, or a 'magic formula', that could be applied to give the correct answer in any given situation. There is a tendency for them to regard such things as a crutch to be used as a substitute for analytical thought. Such students would readily latch on to the $m = 2j - 3$ formula discussed above as a universal panacea for determining the stability (or otherwise) of pin-jointed frames. I've got bad news for such readers: the above formula doesn't always work! (In fairness I should point out that there are other students who delight in finding the exception to the rule – and pointing it out to the lecturer.)

Consider the frame shown in Fig. 11.4(c). It contains nine members and six joints, so $m = 9$ and $j = 6$ and it can thus readily be shown that $m = 2j - 3$ in this case, which suggests that the framework is a perfect frame. In fact, an inspection of the frame shows that this is not, in fact, the case. The left-hand part of the frame is an unbraced square, which is a mechanism and can deform in the same manner as the frame shown in Fig. 11.4 (a). But the right-hand part of the frame has double diagonal cross-bracing, which suggests that it is 'over-stable' and contains redundant members in the same way as the frame shown in Fig. 11.4 (b). So, part of the frame shown in Fig. 11.4 (c) is a mechanism and the other part is statically indeterminate, but this does not make an overall perfect frame, as predicted by the formula!

The lesson to be learned from this is that the formula $m = 2j - 3$ should be regarded as a guide only – it doesn't always work. A given frame should always be inspected to see whether there are any signs of either (a) mechanism or (b) over-stability.

Frames on supports

Up till now in this chapter we have conveniently ignored the fact that, in practice, frames have to be supported. We therefore need to consider the effects of supports on the overall stability of frames.

In Chapter 10 we learned about the three different types of support (roller, pinned and fixed). We also saw that:

- a roller support provides one restraint ($r = 1$);
- a pinned support provides two restraints ($r = 2$);
- a fixed support provides three restraints ($r = 3$).

Reread Chapter 10 if you are unsure about this.

The $m = 2j - 3$ used above is now modified to $m + r = 2j$ where supports are present. As before, m is the number of members and j is the number of joints. The letter r represents the total number of restraints (one for each roller support, two for each pinned support and three for each fixed support).

(1) If $m + r = 2j$, then the frame is a perfect frame and is statically determinate (SD), which means it can be analysed by the methods outlined in the following chapters of this book.

(2) If $m + r < 2j$, then the frame is a mechanism – it is unstable and should not be used as a structure.
(3) If $m + r > 2j$, then the frame contains redundant members and is statically indeterminate (SI), which means it cannot be analysed without resorting to advanced methods of structural analysis.

Examples

For each of the frames shown in Fig. 11.5, use the equation $m + r = 2j$ to determine whether the frame is (a) statically determinate, (b) a mechanism or (c) statically indeterminate. Where the frame is a mechanism, indicate the manner in which the frame could deform. Where the frame is statically determinate, consider which members could be removed without affecting the stability of the structure. The answers are given in Table 11.3.

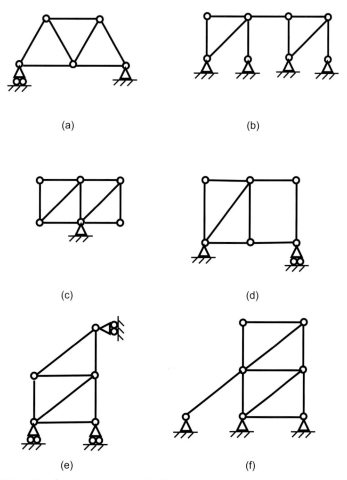

Fig. 11.5 Are these structures stable?

Table 11.3 Stability of structures shown in Fig. 11.5

	m	j	2j	r	m + r	Is m + r = 2j? (or > or <)	Stability type
Figure 11.5 (a)	7	5	10	3	10	=	SD
Figure 11.5 (b)	9	8	16	8	17	>	SI
Figure 11.5 (c)	9	6	12	2	11	<	Mech
Figure 11.5 (d)	8	6	12	3	11	<	Mech
Figure 11.5 (e)	7	5	10	3	10	=	SD
Figure 11.5 (f)	10	7	14	6	16	>	SI

The frames shown in Figs 11.5 (b) and (f) are statically indeterminate. This means they are over-stable and that one or more members may be removed. In the case of Fig. 11.5 (b), one of the diagonal members may be removed (but not both of them!) and the structure would still be stable. In Fig. 11.5 (f), the 'lean-to' diagonal member may be removed without compromising stability. The frames shown in Figs 11.5 (c) and (d) are mechanisms. The structure in Fig. 11.5 (c) is obviously unstable, being free to rotate about its single central support. In Fig. 11.5 (d), the square part of the frame is free to deform in the manner indicated in Fig. 11.4 (a).

Stability of 'real' structures

In practice, the stability of a structure is assured in one of three ways:

(1) Shear walls/stiff core.
(2) Cross-bracing.
(3) Rigid joints.

Let's look at each of these in more detail.

Shear walls/stiff core

This form of stability is usually (but not exclusively) used in concrete buildings. Consider the structural plan of an upper floor of a typical concrete office building, as shown in Fig. 11.6 (a). The structure comprises a grid layout of columns, which support beams and slabs at each floor level. The wind blows horizontally against the building from any direction. It is obviously important that the building doesn't collapse in the manner of a 'house of cards' under the effects of this horizontal wind force. We could design each individual column to resist the wind forces, but for various reasons this is not the way it is normally done.

Instead, **shear walls** are used. These walls are designed to be stiff and strong enough to resist all the lateral forces on the building. Since most buildings have staircases and many have lift shafts, the walls that surround the staircases and lift shafts are often designed and constructed to

A few words about stability   95

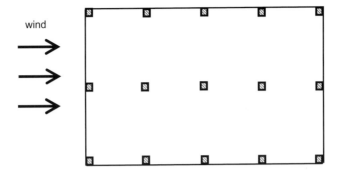

(a) Typical floor plan of reinforced concrete office building

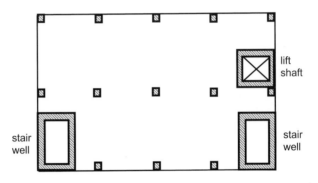

(b) Same floor plan with shear walls added

Fig. 11.6   Provision of stability using shear walls.

perform this role, as shown in Fig. 11.6 (b). On larger buildings, the shear walls may be constructed in such a way as to comprise an inner core to the building, which often contains stairwells, lift shafts, toilets and ducts for services. The NatWest Tower in London is an example of this form of construction.

Cross-bracing

This form of stability is common in steel-framed buildings. Figure 11.7 (a) shows the elevation of a three-storey steel-framed building, on which the wind is blowing. There is nothing to stop the building tilting over and collapsing in the manner indicated by the broken lines in Fig. 11.7 (a).

One way of ensuring stability is to stop the 'squares' in the building elevation from becoming trapeziums. Earlier in this chapter we saw that (a) a triangle is the most basic stable structure and (b) a diagonal member can stop a square from deforming (illustrated in Figs 11.1 (b) and (d) re-

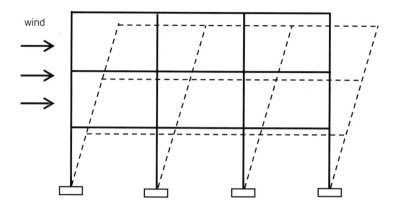

(a) Section through three-storey steel framed building

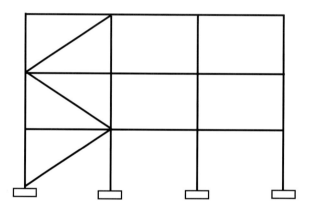

(b) Same section with diagonal bracing added

Fig. 11.7 Provision of stability using cross-bracing.

spectively). So diagonal cross-bracing is used to ensure stability, as shown in Fig. 11.7 (b).

Large modern retail 'sheds', often occupied by do-it-yourself and electronics retailers, are found in most large British towns and cities. These are usually single-storey steel structures and the structure of the building is often visible internally. Next time you visit such a store, have a look at its structure. You will notice steel columns at (typically) 5–6-metre intervals along the building. If you look at the end bay (i.e. the space between the end column and the next one) you may well see a zigzag arrangement of diagonal members. They are there for the reason discussed above: to provide lateral stability to the building as a whole. Figure 11.8 shows quite an 'extrovert' example of diagonal bracing on a new office building; note also the steel truss 'bridge' above the main entrance.

Fig. 11.8 Office building, Euston Road, London.

Rigid joints

A third method of providing lateral stability is simply to make the joints strong and stiff enough that movement of the beams relative to the columns is not possible. The black blobs in Fig. 11.9 indicate stiff joints that stop the action depicted in Fig. 11.7 (a) from happening.

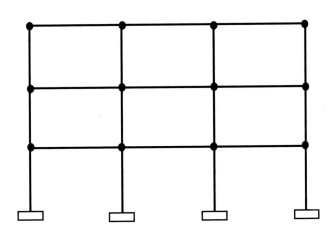

Fig. 11.9 Provision of stability using rigid joints.

What you should remember from this chapter

- All structures must be stable, otherwise they may collapse. Being strong is not sufficient.
- A given structural framework may be either unstable, stable or over-stable. Which of these conditions applies can be determined through a combination of inspection and calculation.
- Lateral stability in buildings can be ensured through one of three means: shear walls, diagonal bracing or rigid connections.

Tutorial examples

(1) For each of the examples shown in Fig. 11.10, determine whether the frame is (a) a perfect frame (SD), (b) unstable (a mechanism) or (c) over-stable (containing redundant members). If the framework is unstable, state where a member could be added to make it stable. If the frame is over-stable, determine which members could be removed and the structure would still be stable.

(2) Select a framed structure near where you live. Determine how lateral stability is provided to the structure and state the reasons why the designer may have chosen that particular method of ensuring stability.

A few words about stability

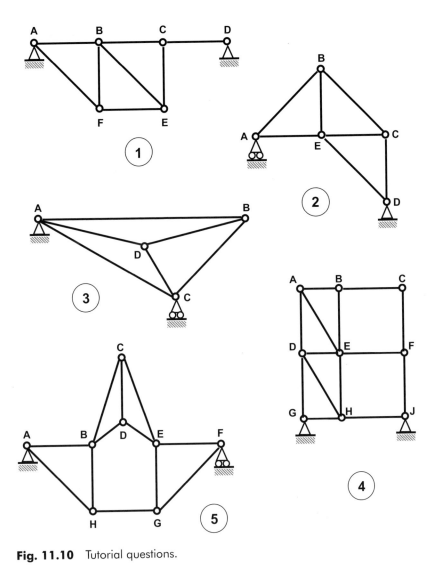

Fig. 11.10 Tutorial questions.

Introduction to the analysis of pin-jointed frames

Simple beams, lattice girders and trusses

The concept of a *beam* has been discussed in earlier chapters. We have seen that if a simple beam in a building is loaded from above, it will sag, as shown in Fig. 12.1. You can readily imagine that the material in the top of the beam is being squashed, or *compressed*. By contrast, the material in the bottom of the beam is being stretched – it is in *tension*.

The amount of downward movement, or *deflection*, from the horizontal depends in part on the material used – it is obviously a lot easier to bend a beam made of rubber than a beam of the same size made of timber!

Another factor that dictates the deflection of a beam is the shape and size of the beam's cross-section. If we consider a beam of rectangular cross-section, the shallower the beam is, the easier it is to bend. The reader can easily verify this point by gripping a plastic ruler at both ends and trying to bend it. If the ruler is orientated with its flat surface horizontal, it is easy to bend in a vertical plane. On the other hand, if the ruler is positioned 'on edge', it is very difficult to bend in a vertical plane, as shown in Fig. 12.2.

We can deduce from this that, all other things being equal, the deeper a beam is, the stronger it is. (This principle is demonstrated mathematically in Chapter 19.)

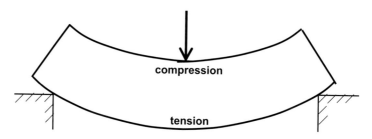

Fig. 12.1 Bending of beams.

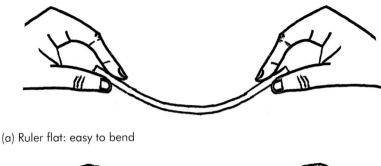

(a) Ruler flat: easy to bend

(b) Ruler on edge: hard to bend

Fig. 12.2 Deeper beams are stronger.

The problem is that, while a deep beam may be stronger than a shallow beam, it also requires more material. Material costs money! You might argue that use of more material is a price worth paying for a stronger beam, but there is a way round it. Instead of having a solid deep beam, it is possible to achieve the same result by having a framework of members, as shown in Fig. 12.3. The top and bottom members (or 'booms' as they are sometimes called) will be, respectively, in compression and tension, just as the top and bottom parts of a solid beam are. Such a framework is called a *lattice girder* or *truss* – it is usually made of steel but can be made of timber. You will have seen railway bridges that look like Fig. 12.3.

Other examples are shown in Figs 12.4–12.6. Figure 12.4 shows a modern lattice footbridge over a river; Fig. 12.5 illustrates a storey-depth lattice truss used in a building structure; and Fig. 12.6 shows the roof of a railway station where several different steel lattice girders are used for support.

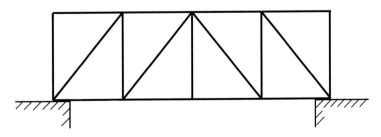

Fig. 12.3 A steel railway bridge.

Fig. 12.4 Trussed bridge across River Spree, Berlin.

Fig. 12.5 Truss in façade, Sony Centre, Berlin.

What is a pin-jointed frame?

Frameworks of structural members, such as steel railway bridges or pylons (as illustrated in Fig. 12.7), are often analysed as pin-jointed frames. This means that the nodes, or joints, between members are regarded as pins, or

Introduction to the analysis of pin-jointed frames **103**

Fig. 12.6 Roof structure, Manchester Victoria station.

Fig. 12.7 Structural steel frameworks.

hinges, which, by definition, cannot transmit moments from one member to another. (See Chapter 10 for an explanation of the concept of a pin.)

It can be shown that the forces in the members of such frameworks are purely axial. In other words, the forces in the members act along the line of the members, which means that each member experiences one of the following:

- pure compression, or
- pure tension, or
- no axial force.

The members of a pin-jointed frame do not experience bending or shear forces.

You may question whether it is legitimate to analyse real structures as pin-jointed frames. After all, if you inspect the junction of two steel members in a railway bridge or electricity pylon you will find that the junction is effected using a combination of angle plates, bolts and welds, and may be quite complex – so surely it can't really be regarded as a pin joint?

Quite right – joints in structural frameworks are not usually pin-jointed in practice. But we consider the joints as pinned, for the purposes of analysis, for the following two reasons:

(1) If you were to analyse the same structure (a) assuming pin joints, then (b) as a rigid-jointed structure, the results would be similar.
(2) It is far easier to analyse the joints as being pins.

It follows therefore that we have to be able to analyse pin-jointed frames.

How are pin-jointed frames analysed?

By the term 'analysis' in the context of pin-jointed frames, we mean calculating:

(1) the *force* in each member;
(2) whether the force is *tensile* or *compressive*.

There are three techniques for doing so:

(1) Method of resolution at joints.
(2) Method of sections.
(3) Graphical method.

These are discussed in Chapters 13, 14 and 15 respectively.

13 Method of resolution at joints

Introduction

The method of resolution at joints is the first of three alternative techniques for analysing pin-jointed frames. By 'analysing' we mean the process of calculating the force in each member of the pin-jointed frame and determining whether each of these forces is in tension or compression.

There are two other techniques:

- method of sections;
- graphical (or force diagram) method.

These two techniques are discussed in Chapters 14 and 15 respectively. The method of sections is appropriate only if the forces in some (rather than all) of the members are required. The graphical method, as its name suggests, involves scale drawing, which by its very nature introduces errors.

Students often have difficulty in understanding the techniques for analysis of pin-jointed frames. This is because these techniques are partly intuitive in nature. Because of these difficulties, students of architecture are often not taught how to analyse pin-jointed frames; if they receive any tuition on pin-jointed frames at all, it is usually merely conceptual. Some lecturers prefer to teach the graphical method to civil engineering students because (a) it is non-mathematical and (b) there is a rigid procedure to be followed, which makes it easier to teach and also easier for students to understand. However, the method of resolution at joints has more universal application and hence it will be taught in this chapter.

The rules

Throughout the analysis of pin-jointed frames using the method of resolution of joints, there are three rules to remember. These rules have all been taught in earlier chapters of this book and are as follows:

106 Basic Structures for Engineers and Architects

Rule No. 1: Force acts in same direction as member

The forces in any member of a pin-jointed frame are axial. In other words, the forces act along the centre line of a member. So, if a member is vertical, the forces in that member must be vertical. If a member is horizontal, the forces in that member will be horizontal. And if a member is inclined at an angle of, say, 30 degrees to the horizontal, the forces within the member will act along that line.

Rule No. 2: Equilibrium applies everywhere

The basic rules of equilibrium apply at all nodes (and in all members) in a pin-jointed frame. This means that the sum of all downward forces on the node exactly equals the sum of all upward forces on the node. It also means that the total force to the left on the node exactly equals the total force to the right. See Chapter 6 if you are unclear on this point.

Rule No. 3: Forces can be split into components

If a force acts at an angle (i.e. it is neither horizontal nor vertical), then that force can be resolved into components – one horizontal and one vertical – which, taken together, have the same effect as the original force. Remember, if a force F acts at an angle θ to the horizontal, its horizontal component will always be $F.\cos\theta$ and its vertical component will always be $F.\sin\theta$. (Remember: 'sign up'.) See Chapter 7 if you need to review the concept of components.

Make sure that you fully understand the above three rules before proceeding, as they will come into play at every step in the following examples.

The general approach

As the term 'method of resolution at joints' implies, the technique involves examining each joint of a framework in turn. The easiest joints to analyse are those at which all forces and members are either horizontal or vertical. This is because there are no diagonal members – whose forces would have to be resolved into vertical and horizontal components – at such joints.

Consider Fig. 13.1 (a), which shows the end part of a framework. No diagonal members radiate from corner B. The joint (or node) at this corner is subjected to a vertical force of 30 kN and a horizontal force of 64 kN as shown.

As the structure is presumably stationary, the rules of equilibrium will apply at the joint.

As total force up = total force down, then the vertical member of this framework (member AB) must experience a 30 kN upward force at point B (to oppose the external 30 kN downward force). Similarly, as total force to the left = total force to the right, then the horizontal member BD must

Method of resolution at joints

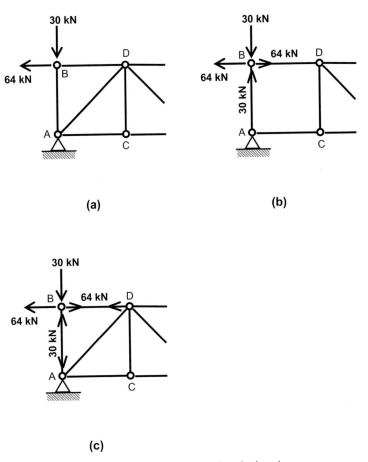

Fig. 13.1 Members in which forces are easily calculated.

experience a 64 kN rightward force at this point (to oppose the external 64 kN leftward force). See Fig. 13.1 (b).

Another thing to remember is that, just as joints must be in equilibrium, so too must members. In the horizontal member, we have a 64 kN force to the right; this must be opposed by a 64 kN force to the left at the other end of the member. In the vertical member there is a 30 kN force upwards; this must be opposed by a 30 kN force downwards at the other end of the member. See Fig. 13.1 (c). In the vertical member the arrows are pointing away from each other, so this member is in *compression*. In the horizontal member the arrows are pointing towards each other, so this member is in *tension* (see Chapter 3 for reminder).

Now look at the framework shown in Fig. 13.2 (a). It would take some time to analyse the whole frame, but there are certain members for which we could determine the forces straight away. Specifically, we could examine the joints at which there are no diagonal members or inclined forces, namely joints B, C and H.

108   Basic Structures for Engineers and Architects

(a)

(b)

Fig. 13.2   More members in which forces are easily calculated.

Using the approach discussed above, we can see straight away that the force in member BD must be 12 kN (to oppose the horizontal 12 kN external force at B) and that it will be in compression (arrows pointing away from each other). Also, the force in member AB must be zero because there is no external vertical force to oppose at point B (or to put it another way, there is an external vertical force of 0 kN to oppose at point B).

Moving on to joint H, we see that the force in member GH must be 24 kN (to oppose the vertical 24 kN external force at H) and that it will be in tension (arrows pointing towards each other). The force in member FH must be zero because there is no external horizontal force at point H to be opposed (or to put it another way, there is an external horizontal force of 0 kN to oppose at point H).

Finally, let's look at joint C. The force in the vertical member CD must be zero because there is no external vertical force to oppose at point C. Furthermore, considering horizontal equilibrium at joint C, the forces in members AC and CE must be equal and opposite – although we cannot obtain their values without further analysis.

The forces we now know are shown in Fig. 13.2 (b).

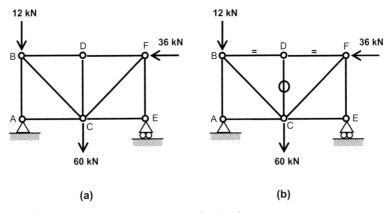

Fig. 13.3 A case which is often misunderstood.

Watch for the catch!

Now look at the frame shown in Fig. 13.3 (a). Looking at the frame, and without doing any calculation, what is the force in member CD?

I have presented this problem to students on numerous occasions. One common answer given to the above question is '60 kN'. I find this depressing because if you think the force in member CD is 60 kN, then I'm afraid you are wrong!

Look at joint D. There is no vertical external force there – or, if you prefer to think of it that way, the vertical external force at D is 0 kN. To balance this, the force in member CD must be 0 kN. The forces in the frame are shown in Fig. 13.3 (b). (Note that, for horizontal equilibrium at joint D, the forces in members BD and DF must be equal and opposite.)

So … why isn't the force in member CD 60 kN?

To answer this question we will look at joint C. Certainly, there is an external downward force of 60 kN there, which, for equilibrium, must be counteracted by an upward force of 60 kN. But member CD will not carry this vertical force alone: diagonal members BC and CF are also present at joint C and will carry a vertical component of force. Therefore the 60 kN upward force is shared between members BC, CD and CF – and, as we saw above, member CD actually carries no force in this case.

Standard cases

From the above discussion we can generate some standard cases of forces in certain members of pin-jointed frames. These standard cases are illustrated in Fig. 13.4.

In Fig. 13.4 (a), consideration of vertical equilibrium at joint A tells us that the force in member AB must be $F1$ to counter the vertical external force of $F1$ at joint A. (Note that the external force of $F3$ at joint B has no direct influence on the force in member AB.) Horizontal equilibrium at joint

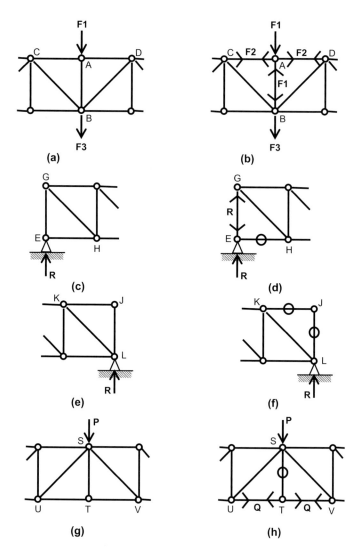

Fig. 13.4 Standard cases.

A tells us that the force in member AC – whatever it is – must be equal to the force in AD and, as the direction of the arrows must oppose each other at joint A for equilibrium, members AC and AD are either both in compression or both in tension. This is illustrated in Fig. 13.4 (b).

As there are no diagonal members present at joint E in Fig. 13.4 (c), the force in vertical member EG must be equal to the support reaction R. Furthermore, the force in horizontal member EH must be zero as there is no opposing horizontal external load. See Fig. 13.4 (d).

If we consider vertical and horizontal equilibrium at joint J in Fig. 13.4 (e) we will see that the forces in members KJ and JL must both be zero as there are neither external forces nor diagonal members at joint J. This is shown in Fig. 13.4(f).

You should realise by now that the force in member ST in Fig. 13.4 (g) is not P. There are diagonal members present at joint S: the vertical components of the forces in these diagonal members will oppose the force P. The force in ST is in fact zero because there is no opposing external vertical force (or diagonal members to provide an opposing vertical force) at joint T. See Fig. 13.4 (h).

Study the standard cases shown in Fig. 13.4 and note particularly the presence or absence of diagonal members at the various joints.

The influence of diagonal members

It would be wonderful, from the analysis point of view, if pin-jointed frames contained no diagonal members. Unfortunately, they always do: diagonal members are required to assure the frame's stability. So how do we analyse joints where diagonal members are present? Look at Fig. 13.5 (a), which shows a joint at the end of a frame. The joint comprises a horizontal member (AB) connected to a member inclined at an angle of 60 degrees to the horizontal (BC). A vertical external force of 3 kN acts at the joint. We wish to find the forces in members AB and BC.

If we resolve vertically at B, we can determine the force in member BC. The total force down at the joint (3 kN) will be equal to the total force up, which must be the vertical component of the force in member BC. So:

$$F_{BC} \cdot \sin 60 = 3 \text{ kN, therefore } F_{BC} = 3.46 \text{ kN}$$

If we now resolve horizontally at B, we can calculate the force in member AB. The force in member AB will be equal to the horizontal component of the force in member BC. So:

$$F_{AB} = F_{BC} \cdot \cos 60, \text{ therefore } F_{AB} = (3.46 \times 0.5) = 1.73 \text{ kN}$$

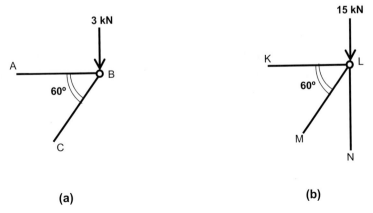

(a)   (b)

Fig. 13.5 Joints with diagonal members.

Now look at the joint L shown in Fig. 13.5 (b). We want to calculate the force in each of members KL, LM and LN, but it is not possible to do so from the information given: if we try to resolve either horizontally or vertically, we generate equations with more than one unknown, which cannot be solved. When analysing a frame with a joint like this, we should not start our analysis at this joint. Instead, we should start at another joint which resembles one of the examples above.

Now we will work through an entire framework in order to calculate all the forces in that frame. (Note: if the above calculation makes no sense at all to you, go back and read Chapter 7 – particularly the part on components.)

Worked example No. 1

See Fig. 13.6. The procedure is as follows:

(1) Calculate the end reactions R_A and R_E in the same way as you would for a beam (see Chapter 9).
(2) Proceed through the framework node by node, using the rules above to calculate the forces (and the directions of those forces) in each member.

Frequently Asked Question: How do I know which node to start at and which order to proceed through the nodes?

This is where the analysis becomes intuitive. You have to start at a node where there is not more than one unknown – but identifying such a node is not easy for the novice. Generally you should start at a support position, then move on to an adjacent node. The following example shows you how.

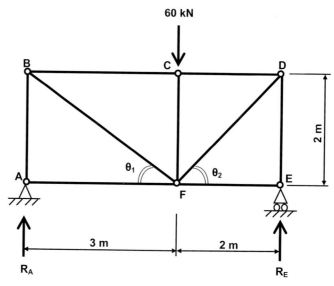

Fig. 13.6 Worked example No. 1.

Determination of reactions

From vertical equilibrium, the total force up ↑ = the total force down ↓.
Therefore

$$R_A + R_E = 60 \text{ kN}$$

This doesn't tell us what R_A is, neither does it tell us what R_E is. It simply tells us that the two of them added together equals 60 kN. To evaluate R_A and R_E, we need another equation. This further equation can be determined from moment equilibrium, discussed in Chapter 6, which tells us that the total clockwise moment about any stationary point is equal to the total anticlockwise moment about that point.

Considering moments about point A

Clockwise moment about point A due to external forces = 60 kN × 3 m

Anticlockwise moment about point A due to external forces = R_E × 5 m

Equating these two:

$$R_E \times 5 \text{ m} = 60 \text{ kN} \times 3 \text{ m}$$

Therefore

$$R_E = 60 \text{ kN} \times 3 \text{ m} / 5 \text{ m} = 36 \text{ kN}$$

Now since $R_A + R_E = 60$ kN (discussed above), then

$$R_A = 60 - 36 = 24 \text{ kN}$$

Applying the 'common sense check' (introduced in Chapter 9): the 60 kN load (which is the only load on the structure) acts to the right of centre, so it will be the right-hand support which 'does the most work' in supporting the structure. Therefore we would expect the right-hand reaction (R_E) to be the greater of the two, which in fact it is (36 kN is greater than 24 kN).

Let's now add the reactions we've calculated to our diagram of the frame. See Fig. 13.7.

Analysis of the frame

Throughout this analysis, the following notation will be used:

F_{AB} represents the force in member AB

F_{BC} represents the force in member BC

... and so on.

Node A

There are three 'legs' to joint A:

- the vertical reaction R_A
- the vertical member AB
- the horizontal member AF.

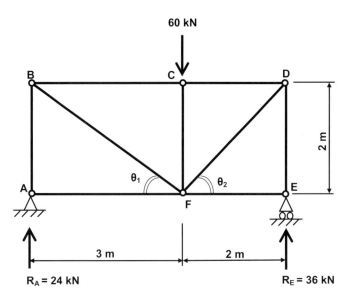

Fig. 13.7 Worked example No. 1 – with reactions calculated.

Resolving vertically at joint A

The term 'resolving vertically' means that we are considering the vertical forces (and vertical components of forces) associated with joint A, mindful of the fact that, for equilibrium, the total upward force at A is equal to the total downward force at A.

Joint A experiences an upward force of 24 kN, in the form of the vertical reaction R_A. This means that, for equilibrium, there must be an (opposing) downward force of 24 kN at A. Since member AF, being horizontal, can contain only a purely horizontal force (i.e. no component of vertical force – see Rule 3 above), the downward 24 kN force can occur only in member AB. Therefore the force in member AB, F_{AB}, is 24 kN and is downwards in direction at end A.

Member AB

The principle of equilibrium applies in all parts of a structure or framework: not only at all nodes but in all members too. We have just determined that the force in member AB is 24 kN downwards at end A. As previously stated, wherever there is a downward force there must be an equal and opposite upward force, so it follows that there must be an upward force of 24 kN in member AB at end B.

Resolving horizontally at joint A

The term 'resolving horizontally' means that we are considering the horizontal forces (and horizontal components of forces) associated with joint A, mindful of the fact that, for equilibrium, the total horizontal force to the left at A is equal to the total horizontal force to the right at A.

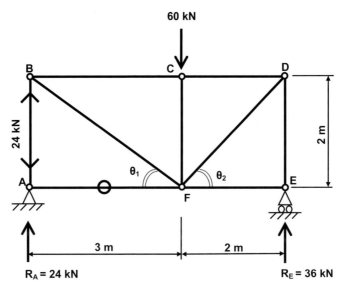

Fig. 13.8 Worked example No. 1 – forces in members AB and AF calculated.

The reaction at A, R_A, is purely vertical and has no horizontal component. Similarly, the force in member AB (which we now know to be 24 kN) is also purely vertical and has no horizontal component. Since there are no other external forces at joint A, the only member at joint A that can experience a horizontal force is member AF. And since there are no other horizontal forces to oppose it, the force in member AF, F_{AF}, must be zero.

Our framework now looks as shown in Fig. 13.8.

We can now carry out a similar analysis of joint E. Using exactly the same approach as we used above for joint A, it can be shown that the force in member DE, F_{DE}, is 36 kN downwards (at end E) and the force in member FE, F_{FE}, is zero.

Our framework now looks as shown in Fig. 13.9.

Joint B
There are three 'legs' to node B:

- the vertical member AB (which contains a vertical force only);
- the horizontal member BC (which contains a horizontal force only);
- the inclined member BF (which, being inclined, will contain both horizontal and vertical components of force).

Resolving vertically at joint B
The only two members connecting at joint B that can have a vertical component of force are AB and BF. (Member BC, being horizontal, has no vertical force – see Rule No. 1 at the beginning of this chapter.)

We already know that there is an upward vertical force of 24 kN at joint B, contained in member AB. For equilibrium, there must be an opposing

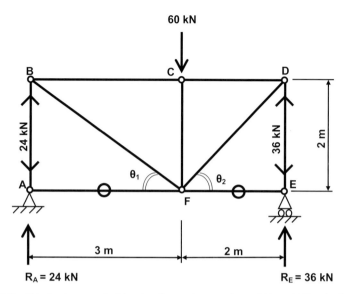

Fig. 13.9 Worked example No. 1 – forces in members DE and EF calculated.

(downward) force of 24 kN and this must occur in member BF (i.e. the only other member at joint B that can contain a vertical force). Therefore the vertical component of the force in member BF must be 24 kN downwards.

Remembering that the vertical component of a force F at an angle θ is F.sin θ, it follows that in this case:

$$F_{BF} \times \sin \theta_1 = 24 \text{ kN}$$

Now θ_1 is the angle AFB = the inverse tan of 2/3 = 33.7°. Therefore

$$F_{BF} \times \sin 33.7° = 24 \text{ kN}$$

So

$$F_{BF} = 24/\sin 33.7° = 43.3 \text{ kN}$$

Let's now consider the direction of this force. We have said that the vertical component of the force in member BF (at end B) must act downwards. This means that the force in member BF (at end B) must act downwards and to the right. Because equilibrium must apply in members as well as joints, this means that the force in member BF at end F must oppose the force at end B; in other words, it must act upwards and to the left.

Because the arrows in member BF point towards each other, member BF must be in tension. (Remember from Chapter 3: if the arrows in a member point towards each other, that member is in *tension*. Think of the letter 'T' – the first letter of the words 'towards' and 'tension'.)

The framework now looks as shown in Fig. 13.10.

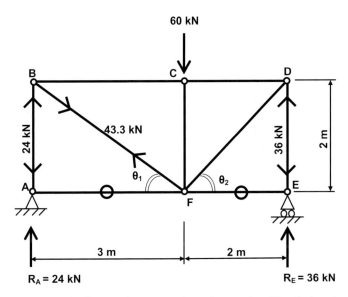

Fig. 13.10 Worked example No. 1 – force in member BF calculated.

Resolving horizontally at joint B
The only two members connecting at joint B that can have a horizontal component of force are BF and BC. (Member BA, being vertical, has no horizontal force – again, see Rule No. 1 at the beginning of this chapter.) If the force in member BF (at end B) is 43.3 kN downwards and to the right, then the horizontal component of this force is $F_{BF} \cos \theta_1 = 43.3 \times \cos 33.7° = 36$ kN (to the right).

For equilibrium, there must be an opposing (to the left) force of 36 kN and this must occur in member BC (i.e. the only other member at joint B that can contain a horizontal force). So, the force in member BC (at end B) is 36 kN to the left. This will be opposed by a force of 36 kN to the right at end C. Therefore the two arrows in member BC point away from each other, so member BC must be in *compression*.

Our framework now looks as shown in Fig. 13.11.

We can now carry out a similar analysis of joint D. Using exactly the same approach as we used above for joint B, it can be shown that the force in member DF, F_{DF}, is 50.9 kN downwards and to the left (at end D) and the force in member DC, F_{DC}, is 36 kN to the right (at end D). (If you don't get those figures, remember that we have a different angle in this case: $\theta_2 = 45°$.)

Our framework now looks as shown in Fig. 13.12.

Joint C
Analysis of joint C is straightforward, as there are no inclined members to complicate matters.

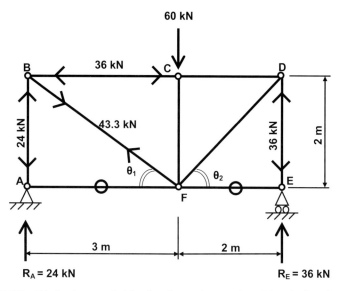

Fig. 13.11 Worked example No. 1 – forces in member BC calculated.

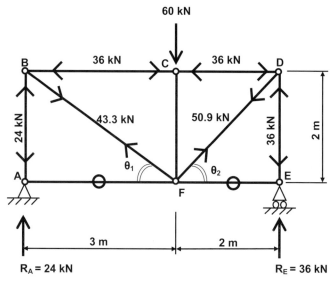

Fig. 13.12 Worked example No. 1 – forces in members DF and DC calculated.

Resolving vertically at joint C

There is a downward external vertical downwards force of 60 kN at joint C. To oppose this, the force in member CF (at end C) must be upwards. The force at the other end of CF will be downwards, therefore the member is in compression.

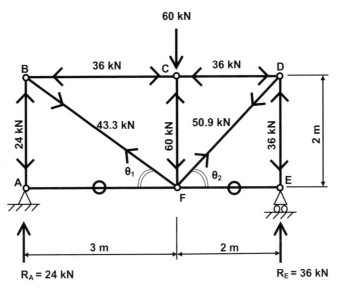

Fig. 13.13 Worked example No. 1 – frame fully analysed.

Resolving horizontally at joint C
The 36kN force in member BC (at end C) is to the right, therefore, to oppose this, the force in member CD (at end C) must also be 36 kN, but to the left. The force at the other end of CD will be to the right, therefore the member is in compression.

The framework now looks as shown in Fig. 13.13.

We have now established the magnitudes and directions of the forces in all the members. So have we finished this example? No, not quite. It would be prudent to carry out a check since, after all, it is quite possible that we may have made a mistake somewhere in our calculations. We can do this by resolving at a point not considered in our earlier calculations and checking, by calculation, that the forces previously calculated balance at that point.

Check: Resolving vertically at joint F
As elsewhere, the total force up at joint F should equal the total force down. There are no external forces at joint F. The following members meet at joint F: AF, BF, CF, DF and EF. Members AF and EF are horizontal so can have no vertical forces (or vertical components of force) in them, so can be ignored when resolving vertically. This leaves members BF, CF and DF.

In our earlier calculations, we found that the vertical components of the forces in members BF and DF are upwards and we found that the vertical force in the (vertical) member CF is downwards. It follows, for equilibrium, that the sum of the vertical components of forces in members BF and DF (acting upwards) must equal the vertical force in member CF (acting downwards).

Vertical component of force in member:

$BF = F_{BF} \cdot \sin \theta_1 = 43.3 \times \sin 33.7° = 24$ kN ↑

Vertical component of force in member:

$DF = F_{DF} \cdot \sin \theta_2 = 50.9 \times \sin 45° = 36$ kN ↑

Vertical force in member:

$CF = 60$ kN ↓

Since 24 + 36 = 60, there is vertical equilibrium at joint F, so our earlier calculations are shown to be correct. A further check could be carried out by considering horizontal equilibrium at joint F.

Worked example No. 2

See Fig. 13.14. A different example, but the principles and the procedure are the same.

Determination of reactions

From vertical equilibrium, the total force up ↑ = the total force down ↓. Therefore

$R_A + R_B = 200$ kN

Once again, this doesn't tell us what R_A is, neither does it tell us what R_B is. It simply tells us that the two of them added together equals 200 kN. To

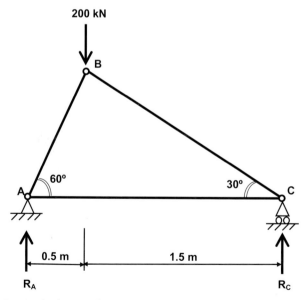

Fig. 13.14 Worked example No. 2.

evaluate R_A and R_B, we need another equation. This further equation can be determined from moment equilibrium, discussed in Chapter 6, which tells us that the total clockwise moment about any stationary point is equal to the total anticlockwise moment about that point.

Considering moments about point A

Clockwise moment about point A due to external forces = 200 kN × 0.5 m

Anticlockwise moment about point A due to external forces = R_B × 2 m

Equating these two:

R_B × 2 m = 200 kN × 0.5 m

Therefore

R_E = 200 kN × 0.5 m/2 m = 50 kN

Now since $R_A + R_B$ = 200 kN (discussed above), then

R_A = 200 − 50 = 150 kN

Applying the 'common sense check': the 200 kN load (which is the only load on the structure) acts to the left of centre, so it will be the left-hand support which 'does the most work' in supporting the structure. Therefore we would expect the left-hand reaction (R_A) to be the greater of the two, which in fact it is (150 kN is greater than 50 kN).

Let's now add the reactions we've calculated to our diagram of the frame. See Fig. 13.15.

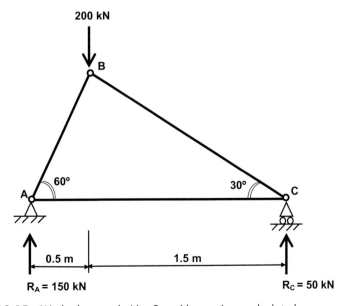

Fig. 13.15 Worked example No. 2 – with reactions calculated.

Analysis of the frame

As before, the following notation will be used throughout this analysis:

F_{AB} represents the force in member AB

F_{BC} represents the force in member BC

... and so on.

Node A

There are three 'legs' to joint A:

- the vertical reaction R_A
- the inclined member AB
- the horizontal member AC.

Joint A experiences an upward force of 150 kN, in the form of the vertical reaction R_A. This means that, for equilibrium, there must be an (opposing) downward force of 150 kN at A. Since member AC, being horizontal, can contain only a purely horizontal force (i.e. no component of vertical force – see Rule 3 above), then the downward 150 kN force can occur only in member AB. Therefore the vertical component of the force in member AB is 150 kN. So

$$F_{AB} \times \sin 60° = 150 \text{ kN}$$

Therefore

$$F_{AB} = 150/\sin 60° = 173.2 \text{ kN}$$

which is downwards (and to the left) in direction at end A.

Member AB

As in the previous example, the downward (and to the left) force of 173.2 kN at end A of member AB must be opposed by an equal and opposite upward (and to the right) force of 173.2 kN at end B. (As the two arrows point away from each other, the member AB is in compression.)

Resolving horizontally at joint A

The term 'resolving horizontally' means that we are considering the horizontal forces (and horizontal components of forces) associated with joint A, mindful of the fact that, for equilibrium, the total horizontal force to the left at A is equal to the total horizontal force to the right at A.

The reaction at A, R_A, is purely vertical and has no horizontal component. But the force in member AB (which we now know to be 173.2 kN) is inclined and therefore will have a horizontal component. Member AC, being horizontal, will also experience a horizontal force. Since there are no other external forces at joint A, the force in member AC must be equal to the horizontal component of the force in member AB – but opposite in direction. So

$$F_{AC} = F_{AB} \times \cos 60°$$

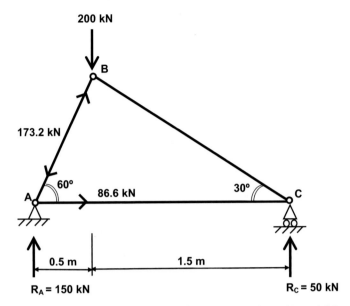

Fig. 13.16 Worked example No. 2 – forces in members AB and AC calculated.

But

$F_{AB} = 173.2$ kN (calculated above)

Therefore

$F_{AC} = 173.2 \times \cos 60° = 173.2 \times 0.5 = 86.6$ kN

Since the horizontal component of the force in member AB (at end A) acts to the left, the horizontal force in member AC (at end A) must act to the right.

Our framework now looks as shown in Fig. 13.16.

We can now carry out a similar analysis of joint C. Using exactly the same approach as we used above for joint A, it can be shown that the force in member CB, $F_{CB'}$ is 100 kN downwards and to the right (at end C) and the force in member BA, $F_{BA'}$ is 86.6 kN to the left (at end C) which, as we would expect, exactly counteracts the force of 86.6 kN to the right at end A of that member. (Since the arrows in member AB point towards each other, the member is in *tension* – remember the letter 'T').

Our framework now looks as shown in Fig. 13.17.

We have now established the magnitudes and directions of the forces in all the members, but, as with the previous example, it would be wise to carry out a check by resolving at a point not considered in our earlier calculations and checking, by calculation, that the forces previously calculated balance at that point. If you were to resolve vertically at joint C, you should find that the forces at that joint balance.

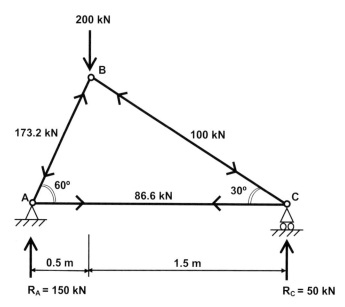

Fig. 13.17 Worked example No. 2 – frame fully analysed.

Tutorial examples

Use the method of resolution at joints to find the forces in all the members of each frame given in Fig. 13.18.

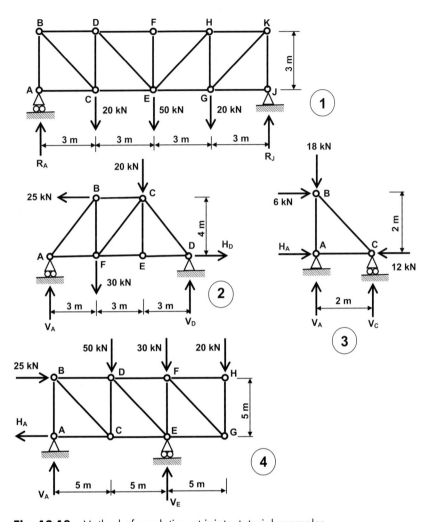

Fig. 13.18 Method of resolution at joints: tutorial examples.

Method of sections

Introduction

Sometimes we don't need (or wish) to determine the axial force in every member of a given pin-jointed frame, as we did when applying the method of resolution at joints in Chapter 13. We may wish to calculate the force in only one or two of the members. In such cases, the method of sections is useful.

In the method of resolution at joints, we doggedly worked our way through the structure, joint by joint, from one end of the structure to another. As you will have found, this can get tedious, particularly when the structure has a large number of members and joints. In the method of sections, we establish a strategically placed 'cut line' through the structure. But determining the correct position of the cut line that will enable us to quickly solve the problem is crucial and partly intuitive, as we shall see.

Figure 14.1 shows another case of one structure – a spherical planetarium building – being encased in another: a huge glass cube supported internally by steel lattice trusses.

Figure 14.2 shows London's Swiss Re building. Popularly known as the 'Gherkin' because of its distinctive shape, it was designed to provide maximum floor space with aerodynamic streamlining. Architect Sir Norman Foster and engineer Ove Arup used an external 'diagrid' (steel members forming a series of triangles) to create the complex curved shape of the building.

Fig. 14.1 New York Planetarium.

Fig. 14.2 Swiss Re building, London.

Background to the method of sections

Imagine a steel framework that forms part of a steel railway bridge, as shown in Fig. 14.3 (a). Let's suppose that we wish to find the axial forces in members AB, BC and CD only. If the railway bridge was an existing structure and we were irresponsible enough to use suitable cutting tools to physically cut through the structure along a line through members AB, BC and CD, as shown in Fig. 14.3 (b), then what would happen? Obviously, the bridge would collapse.

Are there any circumstances under which the bridge would not collapse if cut through as shown? Well, collapse looks pretty inevitable, but there is one circumstance under which (in theory, at least) the bridge would not collapse, as follows:

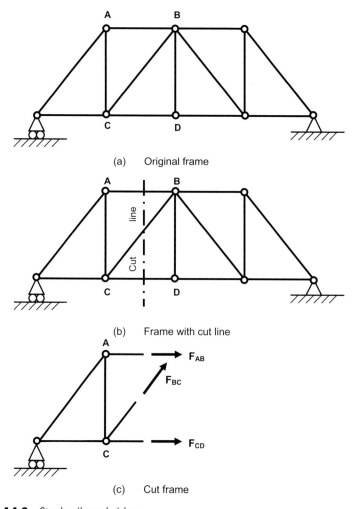

Fig. 14.3 Steel railway bridge.

- If it were possible to use some system of steel ropes, pulleys and props to provide exactly the same forces as existed in the members before they were cut, then the bridge would not collapse.

This means that if we could calculate the external forces in the cut structure that would keep that cut structure in overall equilibrium (indicated as F_{AB}, F_{BC} and F_{CD} in Fig. 14.3 (c)), these would be the same as the internal forces that existed in members AB, BC and CD respectively before they were cut.

To summarise then, the method of sections involves calculating the forces in certain members in a structure by pretending that the members concerned have been cut through and then calculating the external forces on the 'cut' structure. This process will be illustrated through the example that follows.

Example of method of sections

Suppose we wish to calculate the forces in members CD, HD and HG of the structure shown in Fig. 14.4 (a). We need to choose an appropriate cut line. In this case a good choice would be a vertical cut line that passes through all three members, as shown in Fig. 14.4 (a).

First of all, we need to calculate the reactions in the usual way.

Calculation of reactions

From *horizontal equilibrium* of the whole structure:

$H_F = 15$ kN (i.e. Total force $\rightarrow$ = Total force $\leftarrow$)

From *vertical equilibrium*:

$V_A + V_F = 50 + 20 = 70$ kN (i.e. Total force $\uparrow$ = Total force $\downarrow$)

Taking moments about point A (i.e. Total clockwise moment = Total anti-clockwise moment):

$(50 \text{ kN} \times 6 \text{ m}) + (20 \text{ kN} \times 9 \text{ m}) = (V_F \times 12 \text{ m}) + (15 \text{ kN} \times 4 \text{ m}) + (15 \text{ kN} \times 4 \text{ m})$

So

$V_F = 30$ kN

Taking moments about point F:

$(V_A \times 12 \text{ m}) = (15 \text{ kN} \times 8 \text{ m}) + (50 \text{ kN} \times 6 \text{ m}) + (20 \text{ kN} \times 3 \text{ m})$

So

$V_A = 40$ kN

If you don't follow the calculations above then I suggest you revisit the chapters on moments and reactions (Chapters 8 and 9).

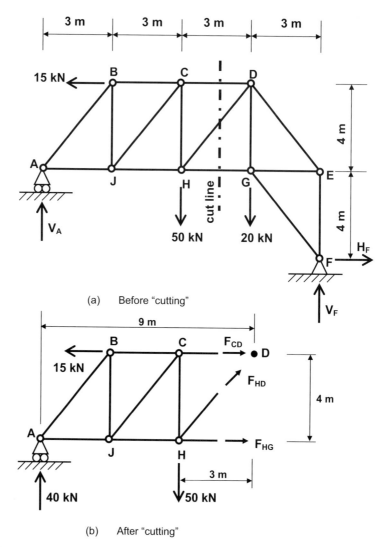

Fig. 14.4 Method of sections example.

The 'cut' section

Let us now suppose that we have cut the frame along the cut line shown in Fig. 14.4 (a). We will discard the part of the frame that is situated to the right of the cut line and will consider only the part to the left, as shown in Fig. 14.4 (b). If we can find the external forces F_{CD}, F_{HD} and F_{HG} that will keep this frame in equilibrium, these forces will correspond to the internal forces that existed in members CD, HD and HG (respectively) in the original pin-jointed frame.

Method of sections

Equilibrium of the frame shown in Fig. 14.4 (b)
Considering *vertical equilibrium*:

$$40 \text{ kN} - 50 \text{ kN} + (F_{HD} \times \sin \theta) = 0 \text{ (i.e. Total force} \uparrow = \text{Total force} \downarrow)$$

You should realise that $(F_{HD} \times \sin \theta)$ is the vertical component of the force in member HD. Revisit Chapter 7 if you are unsure about this.

From basic trigonometry related to a right-angled triangle,

$$\tan \theta = \frac{4 \text{ m}}{3 \text{ m}} = 1.333$$

Therefore $\theta = 53.1$ degrees.
So if

$$40 \text{ kN} - 50 \text{ kN} + (F_{HD} \times \sin 53.1) = 0$$

then

$$F_{HD} = 12.5 \text{ kN}$$

We still need to find F_{CD} and F_{HG}. Let's take moments about point H. (Because the unknown force F_{HG} passes straight through point H, there will be no term involving F_{HG} in the equation if we use H as our 'pivot point' for taking moments. For the same reason, F_{HD} and the vertical 50 kN force at H will not come into the equation either.)

Taking moments about point H
(i.e. Total clockwise moment = Total anticlockwise moment)

$$(F_{CD} \times 4 \text{ m}) + (40 \text{ kN} \times 6 \text{ m}) = (15 \text{ kN} \times 4 \text{ m})$$

So

$$F_{CD} = -45 \text{ kN}$$

(The minus sign indicates that the force acts in the opposite direction to that assumed – so it acts to the left.)

The only remaining force to find is F_{HG}. Although we now know F_{CD} and F_{HD}, it would make life easier if we could take moments about the point through which both of these forces pass (i.e. point D) so there will be no term involving F_{CD} or F_{HD} (or, as it turns out, the 15 kN horizontal force at B). Note that it does not matter that point D is outside the frame we're considering: the rules of equilibrium hold for moments taken about any point, anywhere.

Taking moments about point D
(i.e. Total clockwise moment = Total anticlockwise moment)

$$(40 \text{ kN} \times 9 \text{ m}) = (50 \text{ kN} \times 3 \text{ m}) + (F_{HG} \times 4 \text{ m})$$

So

$$F_{HG} = 52.5 \text{ kN}$$

We've now calculated forces F_{CD}, F_{HD} and F_{HG}. We could check our calculations by considering *horizontal equilibrium* (i.e. Total force → = Total force ←) of the structure shown in Fig. 14.4 (b). But I'll leave that check to you …

So to summarise:

- The force in member CD is 45 kN and is compressive.
- The force in member HD is 12.5 kN and is tensile.
- The force in member HG is 52.5 kN and is tensile.

Summary of the method of sections

(1) Calculate the end reactions in the usual way.
(2) Decide in which member(s) you need to determine the force.
(3) Draw a cut line that cuts through the member(s) of interest. (The cut line may be vertical, horizontal or inclined. It may be necessary to use different cut lines for different members.)
(4) From now on, consider the part of the frame on one side of the cut line only (it doesn't matter which side).
(5) Use the rules of equilibrium to determine the (now external) forces in the members of interest. Consider horizontal and/or vertical equilibrium and take moments about a strategically chosen point. These external forces correspond to the internal forces that existed in the members before they were 'cut'.

What you should remember from this chapter

This chapter outlines the method of sections. This is a useful procedure when we are interested in calculating the forces in only some (e.g. one or two) of the members of a pin-jointed frame. The concept involves pretending that the structure has been cut through the member concerned, then calculating the external forces that would be required to keep the 'cut' structure standing (i.e. in equilibrium). These external forces correspond to the external forces that existed in the 'cut' members before they were cut.

Tutorial examples

Use the method of sections to calculate the axial force and its sense (tension or compression) in the members stated below for each of the pin-jointed frames shown in Fig. 14.5:

- Frame No. 1: CD, DE, EG and GH.
- Frame No. 2: BE and BF.
- Frame No. 3: BC, CD and DE.

Check your answers using the method of resolution at joints (Chapter 13).

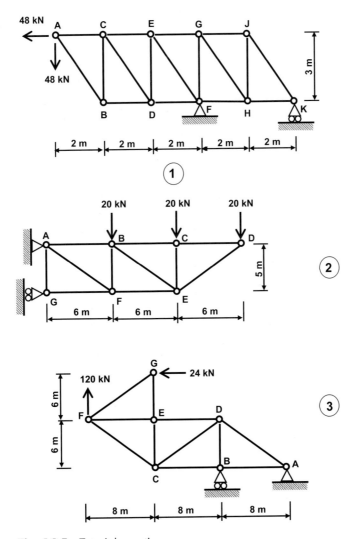

Fig. 14.5 Tutorial questions.

Tutorial answers

(All units are in kN)

- Frame No. 1: 57.6 (c), 48 (T), 144 (T), 129.8 (T).
- Frame No. 2: 62.5 (T), 60 (C).
- Frame No. 3: 296 (T), 200 (C), 112 (C).

Graphical method

Introduction

The previous two chapters discussed two methods of analysing pin-jointed frames, namely the method of resolution at joints and the method of sections. Both of these techniques are mathematical in nature, involving calculation. There is a third technique, called the graphical method (also known as the force diagram method). The graphical method is the subject of this chapter.

The graphical method involves no mathematical calculation whatsoever once the reactions have been calculated in the usual way. This, in itself, makes it appealing to some students. As the name suggests, the member forces and the type of forces are determined by constructing scale diagrams, for which you will need graph paper.

Example 15.1

The graphical method is best explained through example. The example we shall be working through in this chapter is illustrated in Fig. 15.1.

In the previous methods for pin-jointed frame analysis we labelled the joints. In the graphical method, we don't label the joints; instead, we label the areas or zones between the members of the frame and we do so in accordance with Bow's Notation, which is outlined below.

The graphical (force diagram) method in brief

(1) Draw a load line for the applied loads and reactions, to scale. Start from the left-hand support.
(2) Using Bow's Notation (see below), construct a force diagram, one joint at a time, drawing each line parallel to the direction of the member in the framework.

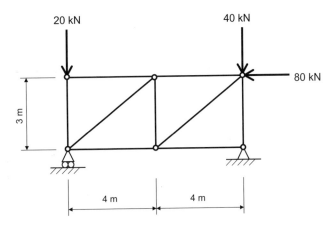

Fig. 15.1 Graphical method example.

(3) Load values can now be scaled from the diagram.
(4) To determine the type of load in a member (i.e. tensile/compressive), 'travel' clockwise round a joint and note the force direction at the joint.
(5) Construct a table.

Bow's Notation

Bow's Notation, named after its creator, is a convention for labelling the various zones in a diagram of a pin-jointed frame. Bow's Notation suggests the following:

(1) Letter the spaces between the external applied loads and reactions.
(2) Number the spaces between internal members.
(3) Start with the letter 'A' between the reactions and work round the frame in a clockwise direction.
(4) Start with the number 1 in the first left-hand space inside the framework.

If we label our frame in accordance with Bow's Notation, it will appear as shown in Fig. 15.2. Notice that the boundaries between the external zones (A, B, etc.) are defined by the positions of the lines of the external forces and reactions and the members of the framework define the frontiers between the internal zones (1, 2, etc.).

As we progress through this problem, we will be constructing a diagram (called a *force diagram*) on a blank piece of graph paper. As we do this we will continually be referring back to the diagram shown in Fig. 15.2, which I will call the frame diagram.

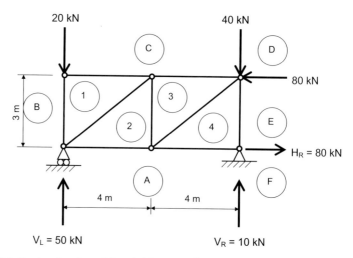

Fig. 15.2 Application of Bow's Notation (frame diagram).

Calculation of reactions

Let's start by calculating the reactions, which we will call V_L (vertical reaction, left-hand support), V_R (vertical reaction at right-hand support) and H_R (horizontal reaction, right-hand support).

From horizontal equilibrium,

$$H_R = 80 \text{ kN} \rightarrow$$

From vertical equilibrium,

$$V_L + V_R = 20 + 40 = 60 \text{ kN}$$

Taking moments about left-hand support:

$$(40 \text{ kN} \times 8 \text{ m}) = (80 \text{ kN} \times 3 \text{ m}) + (V_R \times 8 \text{ m})$$

So

$$V_R = 10 \text{ kN}$$

and therefore

$$V_L = 50 \text{ kN}$$

Construction of the force diagram

We are now in a position to start constructing the force diagram. The various stages in the construction of this diagram are illustrated in Fig. 15.3.

Start with a blank piece of graph paper. Somewhere in the middle of the sheet, select a point and label it *a*. This (lower-case) *a* symbol on the force diagram corresponds to the (upper-case) zone A on the frame diagram. On the frame diagram (Fig. 15.2), you will notice that to get from zone A to zone B you need to cross a 50 kN upward force. This is represented on the

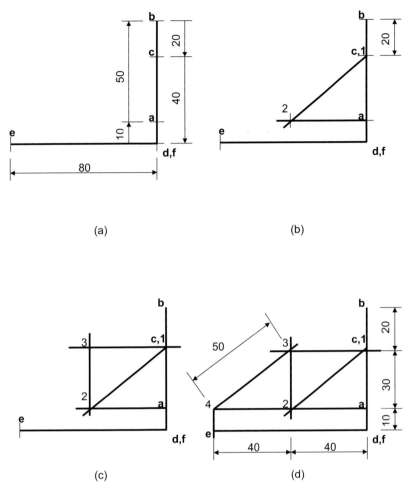

Fig. 15.3 Force diagram.

force diagram by drawing a line vertically upwards from position a (representing zone A) for a distance representing 50 kN to arrive at a new position b (which represents zone B). To do this on graph paper, you will need to adopt a suitable scale – I would suggest a scale of 1 mm = 1 kN would be suitable for this problem on an A4 sheet of graph paper.

So … the line 50 mm long, going up from point a to point b on the force diagram, represents the upward force (reaction) of 50 kN that you have to cross to get from zone A to zone B on the force diagram.

Returning to the frame diagram, getting from zone B to zone C entails crossing a 20 kN downward force (see Fig. 15.2). On the force diagram (Fig. 15.3) this is represented by drawing a line vertically downwards from position b of length 20mm (equivalent to 20 kN). The point arrived at is labelled c and represents zone C on the frame diagram.

Back with the frame diagram again, it can be seen that:

- getting from zone C to zone D involves crossing a 40 kN force (vertically downwards);
- getting from zone D to zone E involves crossing an 80 kN force (to the left);
- getting from zone E to zone F involves crossing an 80 kN force (to the right); and
- getting from zone F to zone A involves crossing a 10 kN force (vertically upwards).

These are represented, respectively, by:

- a vertical line downwards from c, 40 mm long, to establish d;
- a horizontal line leftwards from d, 80 mm long, to establish e;
- a horizontal line rightwards from e, 80 mm long, to establish f;
- a vertical line upwards from f, 10 mm long, to establish a.

The resultant force diagram is shown in Fig. 15.3 (a).

The next task is to locate the points 1, 2, 3 and 4 on the force diagram, which respectively represent zones 1, 2, 3 and 4 on the frame diagram. Examine zone 1 on the frame diagram (Fig. 15.2). It is separated from zone B by a vertical member and from zone C by a horizontal member. This dictates that on our force diagram:

- point 1 lies on a vertical line that also passes through point b; and
- point 1 lies on a horizontal line that also passes through point c.

So point 1 (representing zone 1) must lie at the point shown on Fig. 15.3 (b).

Moving on to zone 2 on the frame diagram, it can be seen that this is separated from zone A by a horizontal member and from zone 1 by a diagonal line sloping upwards and to the right at an angle of '4 squares along, 3 squares up' (or 36.9 degrees). So point 2 can be found on our force diagram from the following two rules:

- point 2 lies on the diagonal line (angle defined above) that also passes through point 1; and
- point 2 lies on a horizontal line that also passes through point a.

So point 2 must lie at the point shown on Fig. 15.3 (b).

From a similar process, point 3 lies at the point where a vertical line through point 2 intersects a horizontal line through point c (see Fig. 15.3 (c)) and point 4 lies at the point where a vertical line through point e meets a horizontal line through point a. The completed force diagram is shown in Fig. 15.3 (d).

Using the force diagram to determine the magnitude of forces

Now comes the easy bit. To determine the force in a member, you simply scale off the distance between the relevant two points on the force diagram (Fig. 15.3 (d)). For example, to determine the force in the right-hand diagonal member of the framework, which separates zone 3 from zone 4 on

Table 15.1 Member forces in Example 15.1

Member reference	Axial force in member (kN)
B–1	20
C–1	0
1–2	50
A–2	40
2–3	30
C–3	40
3–4	50
A–4	80
E–4	10

The member references represent the zones (as shown in Fig. 15.2) that the member lies between. For example, member B–1 lies between zones B and 1 (i.e. the left-hand vertical member), member 3–4 lies between zones 3 and 4 (i.e. the right-hand inclined member) and so on.

the frame diagram (Fig. 15.2), you need to measure the distance between points 3 and 4 on the force diagram (Fig. 15.3 (d)). This distance is 50 mm and thus the force in the member is 50 kN. (Note: Please do not scale off the diagrams in this book as they are not to the correct scale. But your own force diagram will be.)

Similarly, to determine the force in the central vertical member, which separates zones 2 and 3 on the frame diagram, it is necessary to measure the distance between points 2 and 3 on the force diagram. It can readily be seen from Fig. 15.3 (d) that this distance is 30 mm and thus the force in the member is 30 kN.

If you were to carry out this process for the remaining members, the forces you would obtain are shown in Table 15.1.

We're now half way to solving this problem. We have worked out the magnitudes of the forces in each member. Keep reading to find out how we determine the type of force (tension or compression) in each member.

The van driver analogy

Imagine you are a delivery van driver, based in the town of Mitchellstown. On a particular day, you have to make deliveries to addresses in three different towns: Pennyport, Jackston and Charlesville. It is up to you to decide the order in which you visit the three towns. The highway system linking the three towns to each other and Mitchellstown is shown in Fig. 15.4 (a). From this you can see that the most efficient two options are:

(1) Mitchellstown – Pennyport – Jackston – Charlesville – Mitchellstown (i.e. a clockwise circuit).
(2) Mitchellstown – Charlesville – Jackston – Pennyport – Mitchellstown (i.e. an anticlockwise circuit).

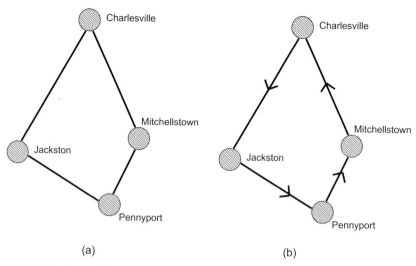

Fig. 15.4 A van driver's delivery route.

You are trying to decide which of the options to go with when your boss comes running out of his office. He tells you he has had an urgent phone call and asks you to do the Charlesville delivery first. So the decision is made for you: you need to visit the towns in the order given in the second option above, shown in Fig. 15.4 (b).

Calculation of the sense (compressive or tensile) of the internal forces in the framework

Returning to the example presented in Fig. 15.1, we have now drawn our force diagram (Fig. 15.3), from which we have scaled off the magnitude of the forces (presented in Table 15.1). But how do we determine which of these forces are in tension and which are in compression?

Joint at top right-hand corner of frame

Consider the top right-hand corner of the frame in our example. By inspection of Fig. 15.2, it can be seen that five zones meet at this point. (If it helps, and if you've got agricultural interests, it might help to consider this point as the place where five fields meet and you have to name them.) The five zones meeting at this point are: C, D, E, 3 and 4.

If we now turned to the force diagram (Fig. 15.3 (d)) and superimposed thick lines on it representing the links between these five points (c, d, e, 3 and 4) we would end up with the diagram shown in Fig. 15.5 (a). Now we know that a 40 kN downward force separates zones C and D (Fig. 15.2), so this can be represented by a downward arrow between points c and d in Fig. 15.5 (a). Similarly, the 80 kN leftward force separating zones D and E can be represented by a leftward arrow between points d and e in Fig. 15.5 (a).

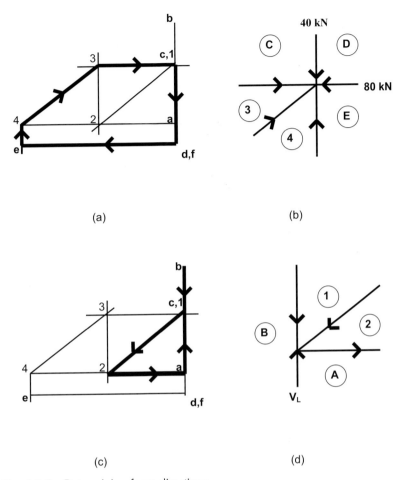

Fig. 15.5 Determining force directions.

By reference to the van driver's analogy above, the directions of these two forces determine the directions of the other forces to complete the circuit in Fig. 15.5 (a) – shown by arrow heads. So the force in line e–4 is upwards, 4–3 is upwards and rightwards and in line 3–c is rightwards, as shown in Fig. 15.5 (a). If we transfer these force directions to the corresponding members of the frame diagram we see that the direction of the forces on the frame diagram will be as shown in Fig. 15.5 (b).

Joint at bottom left-hand corner of frame

Now let's consider the bottom left-hand corner of the frame. Looking at Fig. 15.2, it can be seen that four zones meet at this point, namely A, B, 1 and 2. If we now turned to the force diagram (Fig. 15.3 (d)) and superimposed thick lines on it representing the links between these four points (a, b, 1 and 2) we would end up with the diagram shown in Fig. 15.5 (c).

Now we know that a 50 kN upward force separates zones A and B (see Fig. 15.2), so this can be represented by an upward arrow between points *a* and *b* in Fig. 15.5 (c).

Once again, the direction of this force determines the directions of the other forces to complete the circuit in Fig. 15.5 (c) – shown by arrow heads. This tells us that the force in line *b*–1 is downwards, 1–2 is downwards and leftwards, and 2–*a* is rightwards. So the direction of the forces on the frame diagram will be as shown in Fig. 15.5 (d). Repeating the process for every joint will give the arrow formation shown in Fig. 15.6. Remember:

- Arrows pointing towards each other indicate ***tension***.
- Arrows pointing away from each other indicate ***compression***.

So, to sum up, the procedure for determining which members are in tension and which are in compression is as follows:

(1) Consider each joint in turn.
(2) For the chosen joint, consider which zone numbers/letters directly contact the joint.
(3) Draw a thick line connecting the corresponding zone numbers on the force diagram.

Fig. 15.6 Direction of forces in members.

Fig. 15.7 Steel space frame roof, Lille Europe metro station, France.

(4) The direction of the force between two of the zone numbers is usually known. From this the direction of all the other forces can be determined.

The roof shown in Fig. 15.7, which was photographed from the platform of an underground railway station many metres below, is a typical space frame. A space frame is a three-dimensional pin-jointed frame and has to be designed accordingly.

What you should remember from this chapter

This chapter describes the graphical method, which is a procedure for determining the forces in pin-jointed frames using drawing rather than calculation. The procedure can best be learned by following the example used in this chapter and applying it to the tutorial examples given in the following section.

Tutorial examples

Use the graphical method to determine the forces in each member of each of the examples illustrated in Fig. 15.8. In each case, find out whether the force is tensile or compressive. Then either:

- check your answers using the method of resolution at joints (Chapter 13); or
- check the forces in selected members using the method of sections (Chapter 14).

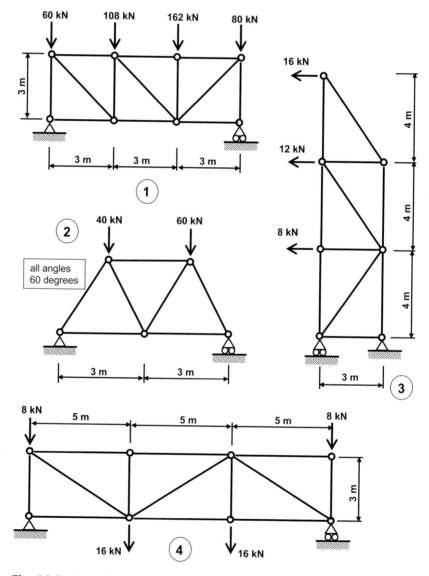

Fig. 15.8 Tutorial questions.

Shear force and bending moments

Introduction

We encountered the concepts of shear and bending in Chapter 3. In this chapter these concepts are explored further and their quantification and calculation are explained.

Deformation of structures

Imagine that the beams indicated by the thick solid horizontal lines in Fig. 16.1 are quite flexible but not particularly strong, so will readily deform under the loads shown. The lines in Fig. 16.2 indicate the deformed (or deflected) forms of the corresponding beams in Fig. 16.1.

Hogging and sagging

We're going to discuss the deformations shown in Fig. 16.2, but before we do so let's define two important terms. You have probably already encountered the term *sagging* – for example, you may have a bed that sags, or dips, in the middle (in which case, my advice is: get a better bed – it's well worth the investment). Sagging, or downward deformation, is illustrated in Fig. 16.3 (a).

Hogging – an upward deformation – is the opposite (or mirror image) of sagging. The concept of hogging is illustrated in Fig. 16.3 (b).

Discussion of the deflected forms shown in Fig. 16.2

Consider, as an example, beam number 1 in Fig. 16.1, which is simply supported at either end and is subjected to a central point load. Clearly, the beam will tend to sag under that load, as indicated by the line in the corresponding diagram in Fig. 16.2. When the beam has sagged, the fibres in the

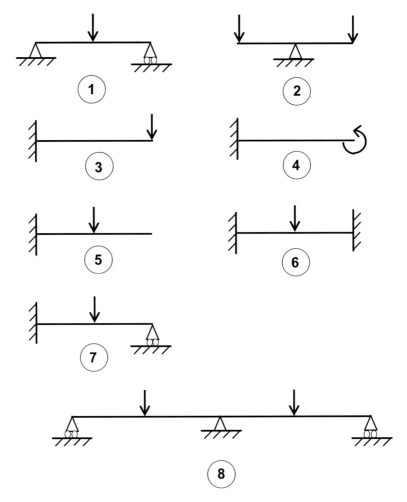

Fig. 16.1 Deformations in beams.

very top of the beam will be squashed together; in other words, they will be compressed. Similarly, the fibres in the bottom part of the beam will have stretched, which indicates that the bottom of the beam is in tension. The fact that the bottom of the beam is in tension is indicated by the letter T (for tension) placed underneath the line in beam number 1 in Fig. 16.2.

Beam number 2 in Fig. 16.1 will tend to hog (or 'break its back') over the central support as a result of the point loads at either end. This hogging profile is indicated by the line in the corresponding diagram in Fig. 16.2. In this case, we will see that the top of the beam will be in tension and therefore we've indicated tension (letter T) above the line at the support position.

We can analyse the remaining beams in Fig. 16.1 in a similar fashion and obtain the deformed profiles and tension positions for each one (indicated by the lines and letter T respectively in Fig. 16.2).

Shear force and bending moments **147**

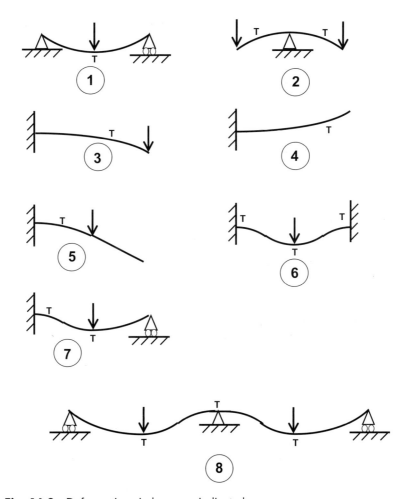

Fig. 16.2 Deformations in beams – indicated.

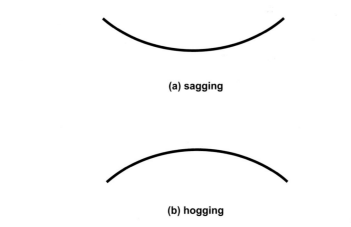

Fig. 16.3 Hogging and sagging.

If you have difficulty visualising the deformation of the beam shown in beam number 4, replicate the situation by holding a standard-length ruler horizontal by gripping it firmly with your left hand at its left-hand end and applying an anticlockwise twist with your right hand at the right-hand end. You will then see the ruler deform in the manner depicted for beam number 4 in Fig. 16.2 and tension will occur on the underside.

When examining the deformed shapes of the beams indicated in Fig. 16.2 for beams 6 and 7, remember that a fixed support firmly grips a beam, while a pinned (or simple) support permits rotation to take place. (See Chapter 10 to remind yourself of the various support types.)

If you completely understand Fig. 16.2, move on to the next section.

Shear and bending

You were introduced to the concepts of *shear* and *bending* in Chapter 3. These two terms represent the ways in which a structural member (for example, a beam) can fail and were illustrated in Figs 3.4 and 3.5. To remind you:

(1) Shear is a cutting or slicing action which causes a beam to simply break or snap. As discussed in Chapter 3, a heavy load located near the support of a weak beam might cause a shear failure to occur.
(2) If a beam is subjected to a load it will bend. The more load that is applied, the more the beam will bend. The more the beam bends, the greater will be the tensile and compressive stresses induced in the beam. Eventually, these stresses will increase beyond the stresses the material can bear and failure will occur – in other words the beam will break. In short, if you increase the bending in a beam, eventually it will break.

So, a beam can fail in shear or it can fail in bending. A natural question at this stage is: which will occur first? Unfortunately, there is no general answer to that question. In some circumstances, a beam will fail in shear; in other cases, a beam will fail in bending. Which happens first depends on the longitudinal profile of the beam: its spans, the position and nature of its supports and the positions and magnitudes of the loading on it. Only by calculation can we tell whether a shear or a bending failure will occur first.

The first thing we need to do is develop a system of *quantifying* shear and bending effects. These quantifications are called *shear force* and *bending moment* respectively and are defined in the following paragraphs.

Shear force

A shear force is the force tending to produce a shear failure at a given point in a beam.

The value of shear force at any point in a beam = the *algebraic sum* of all upward and downward forces to the left of the point. (The term 'algebraic sum' means that upward forces are regarded as being positive and downward forces are considered to be negative.)

> **Example 16.1**
>
> Consider the example shown in Fig. 16.4, in which the end reactions have already been calculated as 25 kN and 15 kN as shown (you should check this). To calculate the shear force at point A, ignore everything to the right of A and examine all the forces that exist to the left of A. Remember, upward forces are positive and downward forces are negative. Adding the forces together:
>
> Shear force at A = + 25 – 30 – 10 = –15 kN
>
>
>
> **Fig. 16.4** Example 16.1: Shear force and bending moment at a point.

Bending moment

The bending moment is the magnitude of the bending effect at any point in a beam. We encountered moments in Chapter 8, where we learned that a moment is a force multiplied by a perpendicular distance, it's either clockwise or anticlockwise and is measured in kN.m or N.mm. The value of bending moment at any point on a beam = the sum of all bending moments to the left of the point. (Regard clockwise moments as being positive and anticlockwise moments as being negative.)

Consider – again – the beam shown in Fig. 16.4. To calculate the bending moment at point A, ignore everything to the right of A and examine the forces (and hence moments) that exist to the left of A. You should realise that, as we are calculating the moment at A, all distances should be measured from point A to the position of the relevant force. See Fig. 16.5 for clarification.

Bending moment at A = (25 kN × 4 m) – (30 kN × 2 m) – (10 kN × 1 m)
= 100 – 60 – 10
= 30 kN.m

Fig. 16.5 Bending moment at point A.

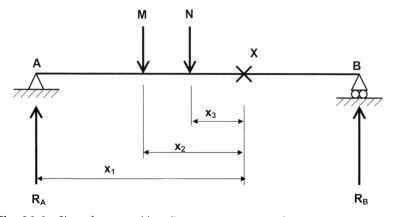

Fig. 16.6 Shear forces and bending moments: general case.

Figure 16.6 shows a more generalised case. Beam AB supports two point loads, M and N, located at the positions shown. The end reactions at A and B are R_A and R_B respectively. Suppose we are interested in finding the shear force at position X, which is located a distance x_1 from the support A, x_2 from point load M and x_3 from point load N. The shear force and bending moment at X are calculated as follows:

Shear force at $X = R_A - M - N$

Bending moment at $X = (R_A \times x_1) - (M \times x_2) - (N \times x_3)$

(Remember: clockwise moments are positive, anticlockwise moments are negative.)

Shear force and bending moment: some examples

In each of the three examples shown in Fig. 16.7, calculate the shear force and bending moment at point D. Check your answers with those given below:

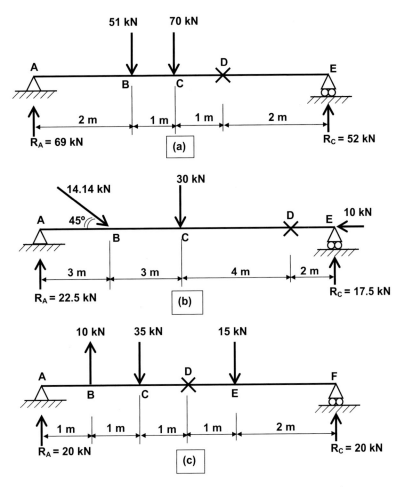

Fig. 16.7 Shear forces and bending moments at a point: examples.

(a) Shear force at D = −52 kN; bending moment at D = 104 kN.m.
(b) Shear force at D = −17.5 kN; bending moment at D = 35 kN.m.
(c) Shear force at D = −5 kN; bending moment at D = 45 kN.m.

(If you are unsure where these answers came from, re-read the examples and rules given above. In example (b), the vertical component of the inclined 14.14 kN force is 10 kN; revisit Chapter 7 for clarification.)

Up till now we've discussed how to calculate values of shear force and bending moment at a specific point in a beam. As engineers and architects though, we're not interested so much in the values at a specific point as in how shear force and bending moment vary along the entire length of a beam. Accordingly, we can calculate and draw graphical representations of shear force and bending moment and their variation along a beam. These are called shear force and bending moment diagrams.

Shear force and bending moment diagrams

Example 16.2

Look at the example shown in Fig. 16.8 (a). The beam is supported at its two ends, A and G, and experiences an 18 kN point load at point E, which is 4 metres from the beam's left-hand end. The reactions at the left and right hand ends are 6 kN and 12 kN respectively, as previously calculated in Chapter 9.

We are going to calculate the shear force and bending moment values at 1 metre intervals along the beam, in other words at points A, B, C, D, E, F and G. When you do this or a similar exercise yourself, I suggest you use graph paper and draw vertical guide lines to make the draughtsmanship easier.

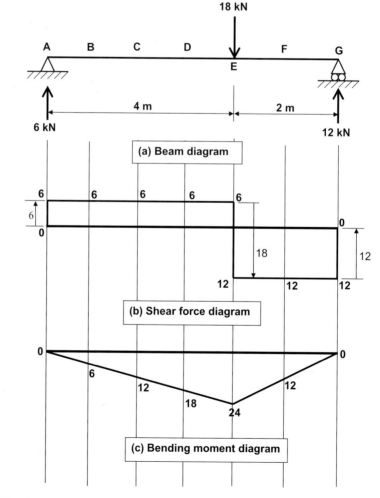

Fig. 16.8 Example 16.2: Shear force and bending moment diagrams.

Shear forces

(Remember, always look at what's going on to the *left* of the point at which you're trying to calculate shear force.) First of all, draw a horizontal straight line representing zero shear force. This will be the base line from which the shear force diagram is drawn.

There is nothing to the left of point A, so the shear force at point A is zero.

If we go a very small distance (say 2 millimetres) to the right of A, there is now a 6 kN upward force to the left of the point we're considering. So the shear force at this point is 6 kN. We can represent this effect by a vertical straight line at point A, starting at the zero force base line and going up to a point representing 6 kN. Each of points B, C, D and E has a 6 kN force to the left of it (i.e. the reaction at point A), so the shear force at each of those points is 6 kN. These values can be plotted on our shear force diagram.

Now consider a point a very small distance (say 2 millimetres) to the right of E. If we examine all the forces to the left of this point, we see that there is an upward force of 6 kN (at A) and a downward force of 18 kN (at E). The shear force at this point must be (6 – 18) = –12 kN (which means 12 kN below the base line). The shear forces at F and just to the left of G will have the same value (–12 kN).

At G itself the sum of all the forces = (6 kN – 18 kN + 12 kN) = 0 kN. So the shear force at G is zero. The shear force diagram is drawn in Fig. 16.8 (b).

Bending moments

Again, we will be looking solely at forces and moments to the left of the point we're considering. We will calculate the moment at each point, remembering that:

- clockwise moments are positive and anticlockwise moments are negative;
- distances are measured from the force concerned to the point considered.

Bending moment at A = +(6 kN × 0 m)   = 0 kN.m

Bending moment at B = +(6 kN × 1 m)   = 6 kN.m

Bending moment at C = +(6 kN × 2 m)   = 12 kN.m

Bending moment at D = +(6 kN × 3 m)   = 18 kN.m

Bending moment at E = +(6 kN × 4 m) – (18 kN × 0 m)   = 24 kN.m

Bending moment at F = +(6 kN × 5 m) – (18 kN × 1 m)   = 12 kN.m

Bending moment at G = +(6 kN × 6 m) – (18 kN × 2 m)   = 0 kN.m

The bending moment diagram is drawn in Fig. 16.8 (c).

Hint: As we're only looking at shear forces and bending moments to the left of a particular point, you might find it helpful, to begin with, to use a piece of paper to cover up the part of the diagram to the right of the point you're considering.

There is an easier way ...

While the above example has given us a good feel for the way in which to calculate and construct shear force and bending moment diagrams, considering every metre along the beam in this way does get rather tedious. A quicker way of drawing the shear force and bending moment diagrams for the above example is as follows.

Shear force diagram – 'follow the arrows'

Draw a base line representing zero shear force. Then start from the left-hand end of the beam. At this point, there is an upward force of 6 kN. So draw a line upwards from the zero line – go up 6 kN, to a value of +6 kN. Going right from A, we encounter no further forces or other features until we reach point E, so the shear force diagram between A and E will be represented by a horizontal straight line between these two points at a value of +6 kN.

At point E there is a downward force of 18 kN. Our shear force diagram will reflect this by dropping down by 18 kN, which takes us from +6 kN to –12 kN. Going right from E, we encounter no further forces or other features until we reach point G, so the shear force diagram between E and G is a horizontal straight line at a value of –12 kN.

At point G there is an upward force of 12 kN. We're already at –12 kN, so the upward force of 12 kN takes us back up to zero. (Note that shear force diagrams *always* end up back on the zero line. If yours doesn't, you've made a mistake somewhere.)

The shear force diagram is shown in Fig. 16.8 (b). Of course, it is the same as calculated before. Note that there is nothing 'magic' about this process. All we've done is follow the arrows. To summarise: if a force goes upwards (for example, the 6 kN reaction at A), then the shear force diagram goes up by that amount. On the other hand, if a force goes downwards (for example, the 18 kN force at E), then the shear force diagram jumps downwards at that point, again by the same amount.

Bending moment diagram – at 'eventful' points only and join the dots

Earlier we calculated the bending moment at 1 metre intervals along the beam. In fact, we need to do this only at 'eventful' points, plot the values and join the dots. 'Eventful' points (my term) are those points where the problem has some feature, e.g. a point load, a reaction or an end of the beam. If in doubt as to whether a particular point is 'eventful' or not, assume that it is. The 'eventful' points on this beam are A, E and G. We previously calculated the moment values at these three points as 0, 24 and 0 kN.m respectively. Plot these values and join the plotted points with straight lines and you have the bending moment diagram shown in Fig. 16.8 (c).

There is one further point to note. You may have been wondering why we elected, in Fig. 16.8 (c), to indicate the bending moment values below the zero line rather than above it. The convention is that the bending moment diagram is plotted on the side of the beam that experiences tension. From the discussion at the beginning of this chapter – and, specifically, from Fig. 16.1 – you will note that in the current example the beam will sag, so tension occurs in the underside of the beam, which suggests that we plot the bending moment diagram below the zero line.

To summarise: the bending moment diagram is drawn either above or below the zero line, dependent on whether the beam experiences tension in the top or bottom at the point concerned (top: above the line, bottom: below the line).

The shape of shear force and bending moment diagrams

If you examine the shape of the shear force and bending moment diagrams above you will notice the following features:

- The shear force diagram is a series of 'steps'; in other words, it contains horizontal and vertical straight lines only.
- The bending moment diagram comprises sloping straight lines.

The above features hold for all cases where a beam is loaded with point loads only (i.e. no uniformly distributed loads).

To summarise: if a beam experiences point loads only, the shear force diagram will be a series of steps and the bending moment diagram will contain only straight lines (usually sloping).

The relationship between shear force and bending moment

You may explore the mathematical relationship between shear force and bending moment at a later stage in your course. One thing to be aware of now is the following rule, which always holds:

Where the shear force is zero, the bending moment is either a local maximum, a local minimum or zero.

If we look again at the example in Fig. 16.8, we see that the shear force diagram touches (or cuts through) the zero line at A, E and G. If we look at the bending moment at each of those three points, we see it is zero at A and G and a maximum (24 kN.m) at E.

This rule is very useful in problems where it is difficult to identify the position of maximum bending moment. In such cases, the key lies in identifying the position(s) of zero shear force.

More examples

Draw the shear force and bending moment diagrams for each of the three beams shown in Fig. 16.7. The solutions are given in Figs 16.18–16.20 at the end of this chapter.

Shear force and bending moment diagrams for uniformly distributed loads

In Chapter 9 we saw how to calculate moments for uniformly distributed loads. You might find it worthwhile to revisit that chapter to refresh your memory. The rule for calculating bending moments for uniformly distributed loads is shown in Fig. 9.5 which, for convenience, is reproduced here as Fig. 16.9. With reference to that figure, the moment of the uniformly distributed load about A is the total load multiplied by the distance from the centre line of the UDL to the point about which we're taking moments. The total UDL is $w \times x$, the distance concerned is a, so:

Moment of UDL about A = wax

Apply this principle whenever you're working with uniformly distributed loads.

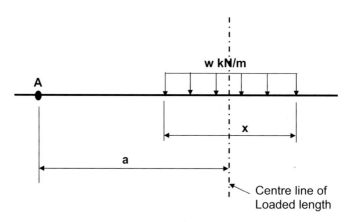

Fig. 16.9 Bending moment calculation for uniformly distributed load (UDL): general case.

Example 16.3

Beam AG, shown in Fig. 16.10, spans 6 metres. It supports a uniformly distributed load of 4 kN/m along its entire length. Draw the shear force and bending moment diagrams.

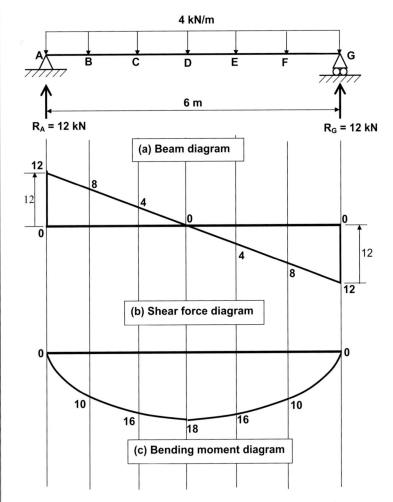

Fig. 16.10 Example 16.3: Shear force and bending moment diagrams: uniformly distributed load example.

First of all, calculate the reactions. This is easy in this case because of the symmetry of both the beam itself and its loading. Each end reaction will be half the total load on the beam. So

$R_A = R_G = (4 \text{ kN/m} \times 6\text{m})/2 = 12 \text{ kN}$

We will now try the metre-by-metre approach – as pioneered in the earlier example – to drawing the shear force and bending moment diagrams. So, we are going to calculate the shear force and bending moment values at points A, B, C, D, E, F and G.

Shear forces

(Remember, always look at what's going on to the *left* of the point at which you're trying to calculate shear force.) As before, draw a horizontal straight line representing zero shear force. This will be the base line from which the shear force diagram is drawn

There is nothing to the left of point A, so the shear force at point A is zero.

If we go a very small distance (say 2 millimetres) to the right of A, there is now a 12 kN upward force (the reaction at A) to the left of the point we're considering. So the shear force at this point is 12 kN. We can represent this effect by a vertical straight line at point A, starting at the zero force base line and going up to a point representing 12 kN.

Each of points B, C, D, E, F and G has this 12 kN upward force to the left of it (i.e. the reaction at point A), but they also have downward forces to the left. Let's consider each of these points in turn.

Point B:
Upward force to left = 12 kN.
Downward force to left = (4 kN/m × 1m) = 4 kN.
Therefore shear force at point B = 12 − 4 = 8 kN.

Point C:
Upward force to left = 12 kN.
Downward force to left = (4 kN/m × 2m) = 8 kN.
Therefore shear force at point C = 12 − 8 = 4 kN.

Point D:
Upward force to left = 12 kN.
Downward force to left = (4 kN/m × 3m) = 12 kN.
Therefore shear force at point D = 12 − 12 = 0 kN.

Point E:
Upward force to left = 12 kN.
Downward force to left = (4 kN/m × 4m) = 16 kN.
Therefore shear force at point E = 12 − 16 = −4 kN.

Point F:
Upward force to left = 12 kN.
Downward force to left = (4 kN/m × 5m) = 20 kN.
Therefore shear force at point F = 12 − 20 = −8 kN.

Immediately left of Point G:
Upward force to left = 12 kN.
Downward force to left = (4 kN/m × 6m) = 24 kN.
Therefore shear force left of point G = 12 − 24 = −12 kN.

At point G, there is an upward reaction of 12 kN. So the net shear force at G will be −12 + 12 = 0 kN.

These values can be plotted on our shear force diagram in Fig. 16.10 (b).

Bending moments

Once more, we will be looking solely at forces and moments to the left of the point we're considering. As in earlier examples, we will calculate the moment at each point, remembering that:

- clockwise moments are positive, and anticlockwise moments are negative;
- distances are measured from the force concerned to the point considered.

Bending moment at A = +(12 kN × 0 m)
= 0 kN.m

Bending moment at B = +(12 kN × 1 m) − (4 kN/m × 1m × 0.5m) = 12 − 2
= 10 kN.m.

Bending moment at C = +(12 kN × 2 m) − (4 kN/m × 2m × 1m) = 24 − 8
= 16 kN.m.

Bending moment at D = +(12 kN × 3 m) − (4 kN/m × 3m × 1.5m) = 36 − 18
= 18 kN.m

Bending moment at E = +(12 kN × 4 m) − (4 kN/m × 4m × 2m) = 48 − 32
= 16 kN.m

Bending moment at F = +(12 kN × 5 m) − (4 kN/m × 5m × 2.5m) = 0 − 50
= 10 kN.m

Bending moment at G = +(12 kN × 6 m) − (4 kN/m × 6m × 3m) = 72 − 72
= 0 kN.m

The bending moment diagram is drawn in Fig. 16.10 (c).

The shape of shear force and bending moment diagrams where uniformly distributed loads are present

If you examine the shape of the shear force and bending moment diagrams in Fig. 16.10 you will notice the following features:

- The shear force diagram comprises sloping straight lines.
- The bending moment diagram is curved (parabolic).

In general, where a beam is loaded with uniformly distributed loads along all or part of its length, the shear force and bending moment diagrams along the part of the beam concerned have the above features.

To summarise: where a beam experiences uniformly distributed loads, the shear force diagram will comprise sloping straight lines and the bending moment diagram will be curved.

Shear force and bending moment diagrams for standard cases

There are three standard cases of beam loading that are so common that the reader would be well advised to commit the results to memory. These are:

- beam with a central point load;
- beam with a non-central point load;
- beam carrying a uniformly distributed load over its entire length.

These cases, along with their respective shear force and bending moment diagrams, are shown in Figs 16.11–16.13. Using the techniques discussed above, you should be able to obtain these reactions and shear force and bending moment values for yourself.

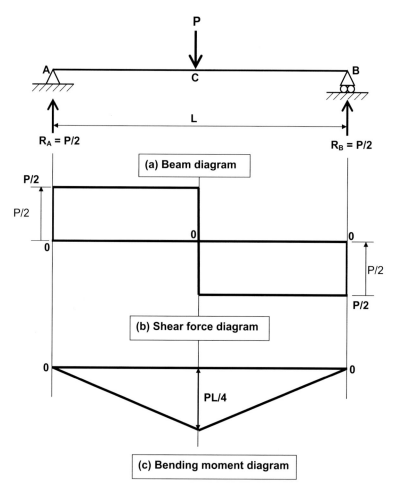

Fig. 16.11 Standard case 1: Shear force and bending moment diagrams for a beam carrying a central point load.

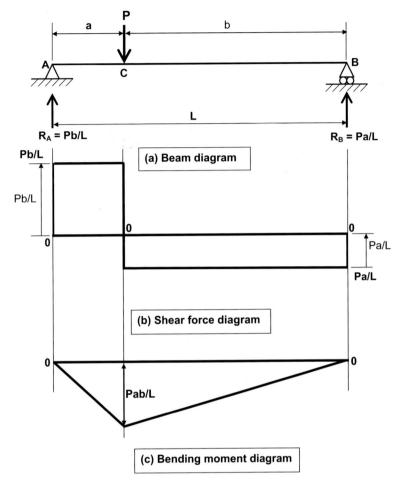

Fig. 16.12 Standard case 2: Shear force and bending moment diagrams for a beam carrying a non-central point load.

Note that the result for the maximum bending moment in a beam with a uniformly distributed load over its entire length ($wL^2/8$) is particularly commonly used in practice.

> Some years ago a colleague of mine in a firm of consulting engineers declared, slightly flippantly: '$wL^2/8$ – that's all you ever need to know!' While this is not quite true (or fair), the comment does at least demonstrate the importance of this result. You might consider that this is underlined by the fact that, in a 'friendly' rafting competition held between the contractor's and resident engineer's staff at a site on which I once worked, the winning raft had been named 'Double You Ell Squared Upon Eight'.

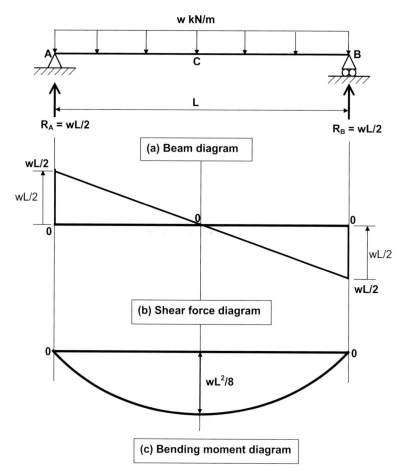

Fig. 16.13 Standard case 3: Shear force and bending moment diagrams for a beam carrying a uniformly distributed load over its entire length.

More examples involving uniformly distributed loads

Draw the shear force and bending moment diagrams for each of the beams shown in Figs 16.14. The solutions are given in Figs 16.21–16.23 at the end of this chapter.

What else can shear force and bending moment diagrams tell us?

Look at the beam shown in Fig. 16.15 (a). It is supported at A and C and experiences a point load at B and at the free end D. By examining the beam and deducing the way in which it might bend (in the same way as we did with the examples at the very beginning of this chapter), we can deduce that:

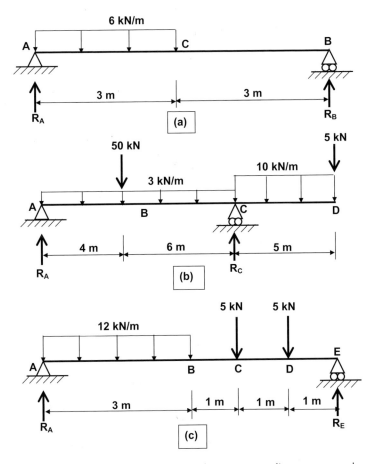

Fig. 16.14 Further shear force and bending moment diagram examples.

- the beam is sagging at point B;
- the beam is hogging at support C;
- the beam is hogging at point D.

Clearly, somewhere between points B and C, the nature of the beam's deflection switches from sagging to hogging. This point is termed the *point of contraflexure*. But where, exactly, does the point of contraflexure occur?

By now you should be able to calculate the reactions and draw the shear force and bending moment diagrams. These are shown in Figs 16.15 (b) and (c) respectively.

Now, earlier in this chapter you were introduced to a convention which stated that the bending moment diagram is always drawn on the tension side of the zero line. This suggests that:

- if the bending moment profile is below the zero line, tension occurs in the bottom face of the beam, which suggests it is *sagging*;

Fig. 16.15 Deflected forms and contraflexure.

- if the bending moment profile is above the zero line, tension occurs in the top face of the beam, which suggests it is *hogging*.

It follows from this that where the bending moment diagram crosses the zero line, the nature of deflection of the beam switches from sagging to hogging (or vice versa). Therefore a point of contraflexure occurs wherever the bending moment profile crosses the zero line. In the current example, that point is 2.5 metres from the left-hand end of the beam. This is determined by recognising that the two (hatched) triangles that constitute the bending moment diagram are *similar* (in the mathematical sense of the word). The deflected profile of the beam is shown in Fig. 16.15 (d).

Example 16.4

Draw the shear force and bending moment diagrams and sketch the deflected form for the beam shown in Fig. 16.16. Identify the position of the points of contraflexure. (The solution is given in Fig. 16.24 at the end of this chapter.)

Fig. 16.16 Example 16.4.

What you should remember from this chapter

- Shear is a cutting or slicing action which causes a beam to break or snap.
- If a beam is subjected to a load it will bend. If the loading is increased, the bending will increase and eventually the beam will break (if it doesn't fail in shear first).
- A shear force is the force tending to produce a shear failure at a given point in a beam.
- The value of shear force at any point in a beam = the algebraic sum of all upward and downward forces to the left of the point.
- A beam will fail in either bending or shear. Which occurs first can only be determined by calculation.
- The bending moment is the magnitude of the bending effect at any point in a beam. The value of bending moment at any point on a beam = the sum of all bending moments to the left of the point.
- Shear force and bending moment diagrams are graphical representations of shear force and bending moment and their variation along a beam.
- The bending moment diagram is drawn either above or below the zero line, dependent on whether the beam experiences tension in the top or bottom at the point concerned (top: above the line, bottom: below the line).
- Where the shear force is zero, the bending moment is either a local maximum, a local minimum or zero. It follows from this that the position of maximum bending moment can be determined from drawing the shear force diagram first.

- If a beam experiences point loads only, the shear force diagram will be a series of steps and the bending moment diagram will contain only straight lines (usually sloping).
- Where a beam experiences uniformly distributed loads, the shear force diagram will comprise sloping straight lines and the bending moment diagram will be curved.
- The point of contraflexure is where the deflected form of a beam switches between hogging and sagging. The bending moment diagram will cross the zero line at this point.
- And don't forget $wL^2/8$!

Tutorial examples

Draw shear force and bending moment diagrams for each of the beams shown in Fig. 16.17.

Fig. 16.17 Further tutorial examples.

Shear force and bending moments

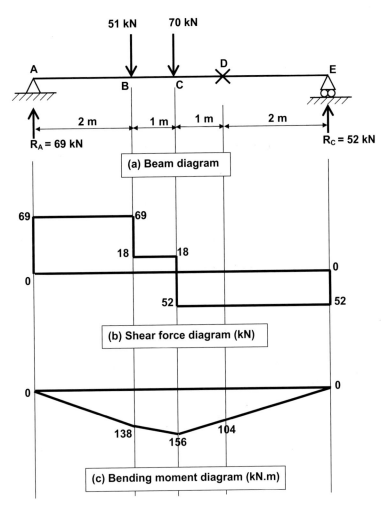

Fig. 16.18 Solution to Fig. 16.7 (a).

168 Basic Structures for Engineers and Architects

Fig. 16.19 Solution to Fig. 16.7 (b).

Fig. 16.20 Solution to Fig. 16.7 (c).

Fig. 16.21 Solution to Fig. 16.14 (a).

Shear force and bending moments

Fig. 16.22 Solution to Fig. 16.14 (b).

172 Basic Structures for Engineers and Architects

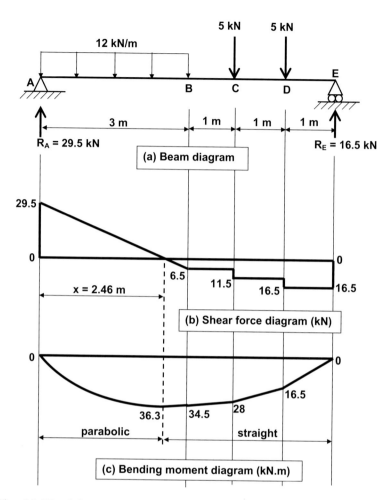

Fig. 16.23 Solution to Fig. 16.14 (c).

Shear force and bending moments

Distance a = 2 m x (40/85) = 0.94 m
Distance b = 2 m x (45/75) = 1.2 m

Fig. 16.24 Solution to Example 16.4.

Figure 16.25 shows a cantilever in a shopping centre in Germany. The cantilever allows a café area on an upper level to overhang (by a modest distance) the pedestrian circulation area below. Note how the depth of the supporting beam reduces towards the 'free' (i.e. unsupported) end. This is because the bending moment in the beam also reduces towards the free end.

Fig. 16.25 Cantilevered balcony in shopping centre.

17 This thing called stress

Introduction

Those of you who are studying to become an architect will be relieved to know that most laymen have some idea of the work that an architect does. However, the civil engineers among you have probably already discovered, to your dismay, that most members of the general public – even the more educated ones – don't have a clue what a civil engineer is or what he or she does, despite the efforts of the relevant professional bodies to promote the profession. However, if pushed, some non-engineers are aware that engineers 'deal with stresses' and that is what this chapter is about.

If some members of the general public are aware of engineers' dealings with stresses, it may have come as a surprise that stresses have hardly been mentioned in the first 16 chapters of this book. However, this and the following three chapters are concerned exclusively with stress.

As we shall see, stress is internal pressure at a point within a structural element occurring as a result of the loads and moments to which the element is subjected. There is a limit to the amount of stress any given material can take, so in structural design it is important to check that this stress is not exceeded.

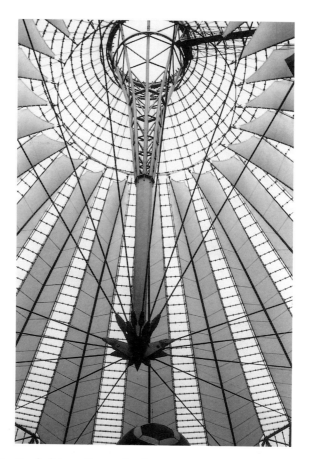

Fig. 17.1 Roof of Sony Centre, Berlin.

Figure 17.1 shows a high-tech tent structure: the roof of the Sony Centre in Berlin.

Fifteen giant steel arches, approximately 7 metres apart, form the main structure of the unusual building shown in Fig. 17.2. The vertical supports to the floors in this building are either suspended off the higher reaches of the arch or, in the case of the end columns, supported off the lower part of the arch. The end support (visible beneath the foliage of the tree in the photograph) is thus in compression and is noticeably fatter than the other vertical supports (which are in tension) because it has to be designed against the possibility of buckling (that is, bending and crumpling).

Fig. 17.2 Ludwig-Erhard-Haus ('The Armadillo'), Berlin.

What is stress?

Stress is internal pressure. Pressure is defined mathematically as Force/Area. As an example, let's suppose you are considering an extended period of foreign travel. As you will be travelling for many months, your extensive preparations will include purchasing a backpack. When fully loaded, the backpack will no doubt be quite heavy, so it's important to select one that will be as comfortable as possible to wear.

From experience, you know that a backpack with narrow shoulder straps will become uncomfortable – if not downright painful – very quickly. An extremely narrow strap – for example, a piece of string – would become extremely uncomfortable and you would probably whimper in agony as the string cut into your shoulders. This is because the load contained in the backpack is transmitted to your shoulder through a comparatively small area, so the pressure will be large (since Pressure = Force/(small) Area).

On the other hand, a broad-strapped backpack will feel much more comfortable. This is because the load from the contents of the backpack will be spread over a much greater area, hence the pressure will be much less. So the message is: choose a backpack with broad shoulder straps and you'll feel much more comfortable.

Whereas pressure is external to an object – e.g. the pressure transmitted into your shoulder through the straps of a backpack or the pressure on a concrete slab due to a heavy piece of machinery or the pressure a building exerts through its foundations to the ground beneath – stress is a similar phenomenon but considered at a point within (for example) a concrete column, a steel beam or a timber joist.

As for pressure, ***direct*** stress is defined mathematically as Force/Area. (Readers should note that it is only direct stress that is thus defined; bending stress and shear stress are different, as we shall see in the chapters that follow.)

Units of stress

As we know, force is measured in Newtons (N) or kiloNewtons (kN) and area is measured in square millimetres (mm²) or square metres (m²). As direct stress is Force/Area, it could be expressed in units of kN/m² or N/mm². In civil engineering we use N/mm² as the units of stress, for the reason that the stresses encountered in practice can be expressed in manageable figures in N/mm² units.

There is a limit to the stress any particular material can take. This stress is known as the ***permissible stress*** or the ***strength*** of the material. Obviously, some materials are stronger than others. For example, the strength of timber is typically in the range 4–7 N/mm², depending on the species. The strength of concrete is typically in the range 25–40 N/mm², while the strength of the steel type normally used in structural steelwork construction is 275 N/mm².

Note the inclination of the main mast of the cable-stayed bridge shown in Fig. 17.3. What does this tell you about the nature of the stresses in the bridge?

Fig. 17.3 Cable-stayed bridge, Bingley bypass, West Yorkshire.

Stress and strain

We use the terms stress and strain in everyday life in circumstances unconnected with structures. For example, you hear people say 'he's under stress' or 'she's feeling the strain'. The 'popular' uses of the words stress and strain are analogous to the technical uses of those words, as we shall see.

Stress can arise as a result of certain situations or circumstances. For example, you might find any of the following situations stressful:

- You are on a plane which is being hijacked.
- Your wife, husband, girlfriend or boyfriend has just announced that s/he is leaving you.
- Your boss tells you he will have to 'let you go'.
- Your car breaks down.

You might react to the stressful situation in a number of ways:

- You might get angry and shout at someone.
- You might burst into tears.
- You might decide that a stiff drink would help.

The *stress* is represented by the situation (the hijacked plane, wife leaving you, etc.) and the *strain* is represented by your reaction to it (the tears, anger or stiff drink).

It's the same principle in structural engineering. For example, a column in a building experiences stress as a result of the forces on it from the floors and walls that the column is supporting. These forces are trying to compress, or squash, the column – in other words, the forces are inflicting *stress* on the column. The column will react to this 'squashing' stress by allowing itself to be reduced in length. This reduction in length (as a proportion of the column's original length) is the *strain*.

Similarly, a hangar cable in a suspension bridge experiences a stress that is trying to stretch the cable, to which it responds by increasing its length. This increase in length (as a proportion of the cable's original length) is the strain.

You will learn more about stress and strain in structural engineering and how to calculate their values in the next chapter.

Fig. 17.4 Concrete arches supporting elevated roadway, Lille, France.

The graceful shallow concrete arches shown in Fig. 17.4 provide an uncluttered public space beneath an elevated roadway.

Types of stress

Direct and shear stresses are discussed in Chapter 18. Bending stress is explained in Chapter 19. Combined bending and axial stresses are investigated in Chapter 20.

What you should remember from this chapter

- Stress is the internal pressure occurring at a given point within a structural element.
- The units of stress are N/mm^2.
- Strain is a measure of what happens as the result of the stress. For example, an extension or reduction in length.

18 Direct (and shear) stress

Introduction

Chapter 17 introduced the concepts of stress and strain. In this chapter we shall discuss direct and shear stresses. We shall also look at how to calculate strains.

Direct (or axial) stress

As discussed in Chapter 17, stress is an internal pressure. A direct (or axial) stress occurs as a result of a direct (or axial) force which acts along the axis of the member (and perpendicular to the member's cross-section). Dependent on the direction of the force, the member may experience tension (causing extension, or stretching, of the member) or compression (which causes contraction, or squashing, of the member). Remember that for equilibrium, forces in one direction must be opposed by equal forces in the opposite direction (see Chapter 6). Examples include a concrete column experiencing a vertical load, as shown in Fig. 18.1 (a) and a steel bar experiencing a horizontal load, shown in Fig. 18.1 (b).

You will notice that the load in the column in Fig. 18.1 (a) is attempting to squash the column, therefore it is inducing a compressive stress in the column. On the other hand, the force on the steel bar in Fig. 18.1 (b) is trying to stretch the bar, so it is producing a tensile stress in the bar. In both cases, if the values of the force (P) and the cross-sectional area (A) are known, the direct (or axial) stress can be calculated using the following equation:

$$\text{Direct stress } (\sigma) = \frac{\text{Force } (P)}{\text{Area } (A)}$$

As explained in Chapter 17, the stress calculated should be expressed in units of N/mm^2.

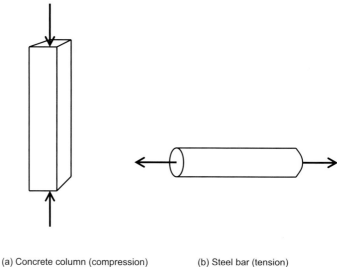

(a) Concrete column (compression) (b) Steel bar (tension)

Fig. 18.1 Direct stresses.

It is important to note that the stress has the same value at every point in the cross-section of the column or bar and it is generally assumed that the stress will be the same throughout the length of the element as well.

Shear stress

A shear stress occurs as a result of a shear force. You will remember from earlier chapters that shear is a cutting or slicing action – for example, if you cut through a loaf of bread with a breadsaw you are applying a shear force to the loaf. Shear forces therefore act perpendicular to the axis of the member. As with direct stresses, shear forces must be opposed by equal forces in the opposite direction – for example, you wouldn't be able to slice a piece of bread without holding the loaf in place with your other hand (which provides the opposing force) at the same time. An example is a timber beam experiencing a shear force – and hence a shear stress – as shown in Fig. 18.2.

If the hatched zone in Fig. 18.2 represents the cross-section (of area A) where shear failure occurs, and the associated shear force is V, the shear stress is calculated from the following equation:

$$\text{Shear stress } (\sigma) = \frac{\text{Shear force } (V)}{\text{Area } (A)}$$

Direct (and shear) stress   **183**

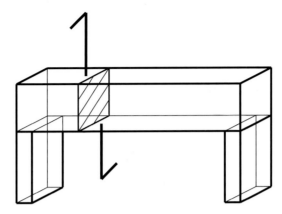

Fig. 18.2 Shear stress in a beam.

As with direct stress, the shear stress calculated should be expressed in units of N/mm².

Note the symbols used for direct stress (σ) and shear stress (τ). These are the standard symbols used in structural engineering. A full list of symbols used in this book is given in Appendix 4.

Strain

The concrete column shown in Fig. 18.1 (a) will reduce in length (by a very small amount, it is hoped) as a result of the compressive axial force to which it is subjected. Similarly, the steel bar in Fig. 18.1 (b) will increase in length (again, by a small amount). These changes in length, as a proportion of the original length of the element, give rise to the *strain*, as defined below:

$$\text{Strain } (\varepsilon) = \frac{\text{Change in length } (\delta L)}{\text{Original length } (L)}$$

It should be pointed out that strain, being simply the ratio of two lengths, has no units. It is a proportion or can, if desired, be expressed as a percentage.

Shear strain is beyond the scope of this book.

We shall now try some numerical examples.

Example 18.1: Stress and strain in compression

A square concrete column in an office building is shown in Fig. 18.3. The column has cross-sectional dimensions 400 mm × 400 mm and supports a total vertical load of 2000 kN. Calculate the direct compressive *stress* at any point in the column.

If the column reduces in length by 3.5 mm as a result of the loading and the column's original length was 4 metres, calculate the *strain* in the column.

Solution
The column is clearly in compression.

The column's cross-sectional area, $A = 400 \times 400 = 160{,}000 \text{ mm}^2$

Axial load $P = 2000 \text{ kN} = 2000 \times 10^3 \text{ N}$

$$\text{Stress } (\sigma) = \frac{\text{Force } (P)}{\text{Area } (A)} = \frac{2000 \times 10^3}{160{,}000} = 12.5 \text{ N/mm}^2$$

$$\text{Strain } (\varepsilon) = \frac{\text{Change in length } (\delta L)}{\text{Original length } (L)} = \frac{3.5 \text{ mm}}{4000 \text{ mm}} = 8.75 \times 10^{-4} = 0.000875$$

(Remember: strain has no units.)

Fig. 18.3 Compressive stress and strain (Example 18.1).

Example 18.2: Stress and strain in tension

The circular steel bar shown in Fig. 18.4 has a diameter of 30 mm and is subjected to a tensile axial force of 50 kN. Calculate the direct tensile *stress* at any point in the bar.

If the bar, whose original length was 2 metres, extends in length by 0.67 mm as a result of the force, calculate the *strain* in the bar.

Solution
The procedure is similar to Example 1, but this time the member is in tension.

The column's cross-sectional area, $A = \pi r^2 = \pi \times 15^2 = 706.9$ mm². (Remember that the radius of a circle is half the diameter. The diameter in this case is 30 mm, so the radius is 15 mm.)

Axial load $P = 50$ kN $= 50 \times 10^3$ N

$$\text{Stress } (\sigma) = \frac{\text{Force }(P)}{\text{Area }(A)} = \frac{50 \times 10^3 \text{ N}}{706.9 \text{ mm}^2} = 70.73 \text{ N/mm}^2$$

$$\text{Strain } (\varepsilon) = \frac{\text{Change in length }(\delta L)}{\text{Original length }(L)} = \frac{0.67 \text{ mm}}{2000 \text{ mm}} = 0.000335$$

Again, strain has no units.

Fig. 18.4 Tensile stress and strain (Example 18.2).

Example 18.3: Shear stress

The shear force at the end of the timber joist shown in Fig. 18.5 is found to be 18 kN. If the timber joist is 50 mm wide and 200 mm deep, calculate the *shear stress* at this point in the joist.

Solution

$$\text{Shear stress } (\sigma) = \frac{\text{Shear force } (V)}{\text{Area } (A)} = \frac{18 \times 10^3 \text{ N}}{50 \times 200 \text{ mm}^2} = 1.8 \text{ N/mm}^2$$

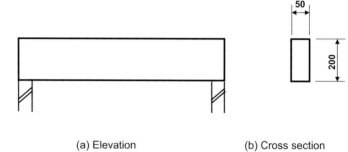

(a) Elevation (b) Cross section

Fig. 18.5 Timber beam in shear.

The relationship between stress and strain

It would be natural at this point to wonder whether or not there is any relationship between stress and strain. We saw in Chapter 17 that strain is a reaction to stress. In Example 18.1 above, we saw that a stress of 12.5 N/mm² in a given concrete column gave rise to a strain of 0.000875. You may wonder whether a stress of double that amount would produce double the strain – or whether tripling the stress would produce triple the strain, and so on. In other words, is stress *proportional* to strain?

If you have studied a materials module you will already know the answer. For most materials, the answer is yes: stress and strain are proportional – up to a point. As you can see from Fig. 18.6, if a graph is plotted of stress versus strain, the graph is a straight line up to a certain point, known as the limit of proportionality. (Beyond the limit of proportionality, the shape of the graph depends on the material but is no longer a straight line.)

If stress is proportional to strain then, mathematically speaking, stress/strain = a constant (Hooke's Law). This constant is known as Young's Modulus, has the symbol E and units of N/mm² or kN/mm².

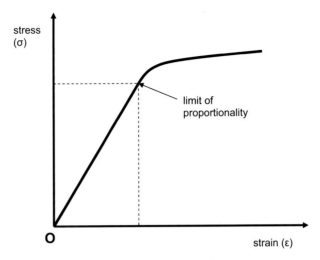

Fig. 18.6 Stress v strain graph.

$$\text{Young's modulus } (E) = \frac{\text{Stress } (\sigma)}{\text{Strain } (\varepsilon)}$$

(For more information on Hooke and Young, see the end of this chapter.)

In Example 18.1 we found that the compressive stress and strain experienced by the concrete column were 12.5 N/mm² and 0.000875 respectively. Therefore:

$$\text{Young's modulus} = \frac{12.5 \text{ N/mm}^2}{0.000875} = 14286 \text{ N/mm}^2 = 14.3 \text{ kN/mm}^2$$

In Example 18.2 we found that the tensile stress and strain experienced by the steel bar were 70.73 N/mm² and 0.000335 respectively. Therefore:

$$\text{Young's modulus for steel} = \frac{70.73 \text{ N/mm}^2}{0.000335} = 211{,}134 \text{ N/mm}^2$$
$$= 211 \text{ kN/mm}^2$$

How to predict change in length

Now we already know that:

$$\text{Direct stress } (\sigma) = \frac{\text{Force } (P)}{\text{Area } (A)}$$

and

$$\text{Strain } (\varepsilon) = \frac{\text{Change in length } (\delta L)}{\text{Original length } (L)}$$

Combining these three equations and rearranging, we get:

$$\text{Change in length } (\delta L) = \frac{PL}{AE}$$

From this equation we can calculate the change in length of a structural element if we know its length (L), the axial load to which it is subjected (P), its cross-sectional area (A) and its Young's Modulus value. The latter can be obtained from scientific data tables if necessary.

Example 18.4 Calculating the change in length of a member under direct stress

A steel tie in a space frame roof structure is originally 2 metres long. If the tie is a solid bar of diameter 40 mm, calculate the extension of the steel bar that would be expected if a tensile force of 150 kN is applied to the bar. The Young's Modulus of steel is 205 kN/mm². If the extension was unacceptably large, what steps could you take to reduce it?

Solution
Cross-sectional area of steel bar = πr^2 = $\pi \times 20^2$ = 1256.6 mm².

$$\text{Change in length } (\delta L) = \frac{PL}{AE} = \frac{150 \times 10^3 \text{ N} \times 2000 \text{ mm}}{1256.6 \text{ mm}^2 \times 205 \times 10^3 \text{ N/mm}^2}$$

$$= 1.16 \text{ mm}$$

This extension of 1.16 mm is small and is probably tolerable in most structures. However, by examination of the 'change in length' formula, the following steps could be taken to reduce the extension if desired:

- Reduce the axial load in the member.
- Reduce the length of the member.
- Increase the cross-sectional area of the member. (This is usually the most practical option.)
- Use a material with a greater Young's Modulus.

What you should remember from this chapter

$$\text{Direct stress } (\sigma) = \frac{\text{Force } (P)}{\text{Area } (A)}$$

$$\text{Shear stress } (\sigma) = \frac{\text{Shear force } (V)}{\text{Area } (A)}$$

$$\text{Strain } (\varepsilon) = \frac{\text{Change in length } (\delta L)}{\text{Original length } (L)}$$

Direct (and shear) stress

$$\text{Young's modulus (E)} = \frac{\text{Stress } (\sigma)}{\text{Strain } (\varepsilon)}$$

$$\text{Change in length } (\delta L) = \frac{PL}{AE}$$

Tutorial examples

(1) Calculate the direct stress in a reinforced concrete column of cross-section 400 mm × 350 mm, subjected to a compressive load of 3000 kN. Express your answer in N/mm² units.
(2) A solid circular steel rod, forming part of a framed structure, is subjected to a tensile force of 750 kN. If the permissible stress in steel is 460 N/mm², what is the minimum diameter of the rod in millimetres? Had the rod been in compression rather than tension, what other factors would need to be considered?
(3) A timber column is subjected to a compressive force of 60 kN. If the permissible compressive stress in timber is 6 N/mm², select a suitable section size for the column. Express your answer in terms of the column's cross-sectional dimensions, in millimetres.
(4) A force is applied to a steel bar, originally 3 metres in length, causing it to extend by 1.5 mm. Calculate the strain (ε) in the bar.
(5) A 3.5 metre long steel tie is subjected to a tensile force of 150 kN. If the bar is round, of diameter 20 mm, and the Young's Modulus (E) value for steel is 200 kN/mm², calculate the change in length of the bar.
(6) An aluminium 'strut' (a compression member) 1.5 metres long is part of a lightweight framed structure and is subjected to a compressive force of 50 kN. Calculate the strain in the strut and determine its change in length. Assume the area of the cross-section is 220 mm² and = 70 kN/mm².
(7) A new suspension bridge in the Far East has one of the longest spans in the world. Each of its main cables is 1 metre in diameter and is designed to sustain an axial tensile force of 13,000 tonnes. Assuming, for simplicity, that each main cable is of solid steel (rather than the collection of many smaller diameter cables that it actually is), calculate the stress in each main cable, in N/mm² units.

Tutorial answers

(1) 21.4 N/mm².
(2) 45.6 mm; buckling.
(3) 100 mm × 100 mm or 75 mm × 150 mm.
(4) $\varepsilon = 0.0005$, or 0.05%.
(5) 8.35 mm.
(6) $\varepsilon = 0.00325$, $\delta L = 4.87$ mm.
(7) $\sigma = 166$ N/mm².

Who were Mr Hooke and Mr Young?

It's claimed that Robert Hooke (1635–1703) was one of the greatest experimental scientists of the 17th century. Certainly he had wide interests in most branches of science and collaborated with other well-known scientists of the day, including Isaac Newton, but unfortunately the two men did not get on. Hooke also had an interest in architecture and he assisted Sir Christopher Wren on the rebuilding of London's St Paul's Cathedral after the Great Fire of 1666. Hooke's Law, discussed in this chapter, is the scientific principle for which he is best remembered.

Thomas Young (1773–1829) also had wide-ranging professional interests. As well as being a physicist whose experiments in elasticity led to the modulus that bears his name, Young was medically qualified and researched extensively in the fields of light and optics.

Bending stress

Introduction

In Chapter 18 we investigated direct stresses – the stresses caused by direct or axial loads on structural elements. In this chapter we will study *bending* stresses. As the name suggests, these are stresses associated with the bending of a beam or other type of structural member.

Bending theory

Consider the beam shown in Fig. 19.1 (a), which is simply supported at its two ends. If a central point load is applied to the beam, it will bend to give the profile shown in Fig. 19.1 (b) Alternatively, if the beam shown in Fig. 19.1 (a) is subjected to a longitudinal load which does not act along the line of the beam's central axis, it will again bend, to give the profile shown in Fig. 19.1 (c).

So, bending can be induced in a beam in one of two ways:

(1) loading perpendicular to the beam's longitudinal axis; or
(2) eccentric axial loads.

If we were to paint vertical stripes at regular intervals along a simply supported beam before loading it, it would appear as shown in Fig. 19.2 (a). After the beam has bent under loading, its profile will resemble Fig. 19.2 (b). You will notice that the stripes in the bent beam shown in Fig. 19.2 (b) are still straight, despite the fact that they are no longer the same distance apart at top and bottom. This would suggest that although the beam has bent, particular cross-sections (as represented by the painted stripes) remain straight and thus have not warped.

Consider the cross-section of a rectangular beam, shown in Fig. 19.3 (a). If the beam bends, we know from our earlier studies that the top part of the beam will be in compression and the bottom part will be in tension.

192 Basic Structures for Engineers and Architects

(a) Beam before load applied

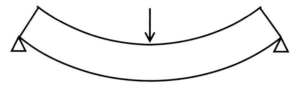

(b) Beam bending caused by central point load

(c) Beam bending caused by eccentric axial load

Fig. 19.1 Bending in beams.

(a) Beam cross sections (edge on) before load applied

(b) Beam cross sections (edge on) after load applied

Fig. 19.2 Effect of bending on beam cross-section.

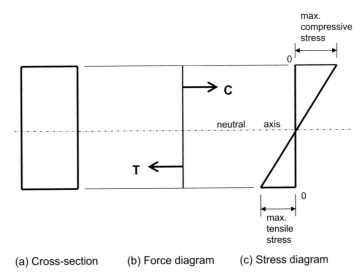

(a) Cross-section   (b) Force diagram   (c) Stress diagram

Fig. 19.3 Bending theory applied to a beam cross-section.

This implies that there must be some level in the cross-section that will be the interface between the compression and tension zones. This interface is called the ***neutral axis*** or ***neutral plane*** and we shall see that there is no stress at this level.

Figure 19.3 (b) is a simple force diagram. The compression in the top part of the beam is represented by force C. The tension in the bottom part of the beam is represented by force T. Note that, as required for equilibrium, forces C and T are equal but opposite in direction.

Figure 19.3 (c) is a stress diagram in which the vertical line represents zero stress. We can readily see that the maximum tension – and hence the maximum tensile stress – occurs at the very bottom of the beam and reduces as we move up the beam from this level. Similarly, the maximum compression – and hence the maximum compressive stress – occurs at the very top of the beam and reduces as we move down the beam. If we join these two maximum values with a straight line, our stress diagram becomes as shown in Fig. 19.3 (c). Note the linear (i.e. straight line) variation in stress as we move down the cross-section.

As we have just seen (Fig. 19.3(c)), tension occurs in the bottom of a beam that is sagging. As concrete is weak in tension, steel reinforcement is provided in the place where it is most useful; that is, near the bottom face of the beam. But site labourers in general, and steelfixers in particular, have not been schooled in structural mechanics. Occasionally you will come across cases where a steelfixer feels it is inconvenient to put all the required steel reinforcement in the bottom face and therefore puts some of it half way up a section. Figure 19.3 demonstrates that any steel placed half way up a section is useless, as the stress is minimal at this point and therefore the steel is not doing any work. The bottom of the section is, accordingly, under-reinforced and therefore likely to fail.

Assumptions for bending theory

(1) The material is *linearly elastic* (as represented by the straight-line graph in Fig. 19.3 (c)).
(2) Young's Modulus (E) is the same in compression and tension. (See Chapter 18 if you need a reminder of Young's Modulus and its significance.)
(3) Material is *homogeneous* (i.e. the same throughout). This is obviously not the case if we're considering a cross-section containing two different materials, e.g. reinforced concrete.
(4) Plane sections remain plane after bending – i.e. no warping. See the discussion of Fig. 19.2 above.

Neutral axis

As discussed above, the neutral axis occurs at the interface of the compression and tension zones of a structural element experiencing bending. The neutral axis has the following characteristics:

- The neutral axis occurs at a level where there is no stress.
- The neutral axis is half way down the cross-section for homogeneous, symmetrical sections.
- The neutral axis passes through the centroid if the material is homogeneous.

The Engineers' Bending Equation

The equation below is known as the Engineers' Bending Equation. The derivation of it is not included here as it contains some fairly scary mathematics, but it can be found in more advanced structures textbooks. It is far more important for you to become familiar with the equation itself – rather than its derivation – and the meaning of the various terms therein:

$$\frac{\sigma}{y} = \frac{M}{I} = \frac{E}{R}$$

where:

σ = bending stress (N/mm²).
i = distance (measured, in millimetres, vertically upwards or downwards) to a particular point from the neutral axis (see Fig. 19.4 (a))
M = bending moment at the point concerned (kN.m or N.mm)
E = Young's Modulus (kN/mm² or N/mm²)
R = radius of curvature (millimetres) (see Fig. 19.4 (b))
I = second moment of area (mm⁴) (see explanation below)

The second moment of area mentioned above is a geometrical property of a cross-section. Its derivation is complex, involving calculus. Suffice it to say

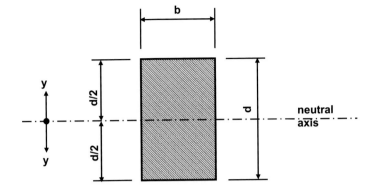

(a) Geometry of a rectangular cross section

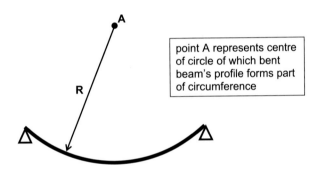

point A represents centre of circle of which bent beam's profile forms part of circumference

(b) Radius of curvature

Fig. 19.4 Engineers' Bending Equation: some terms.

that for a rectangular section of breadth b and depth d, the second moment of area, $I = bd^3/12$.

A further parameter is the **section modulus**, also known as the elastic modulus. This has the symbol z, and is defined as:

$$z = \frac{I}{y_{max}}$$

Now, for a rectangular section, as mentioned above:

$$I = \frac{bd^3}{12}$$

Moreover, for a rectangular section which is homogeneous (same material throughout), the neutral axis must be exactly half way down the section.

Therefore the maximum vertical distance that can be travelled from the neutral axis that still remains within the section is $d/2$. So $y_{max} = d/2$.

So, substituting in the above equation for the special case of a rectangular section:

$$z = \frac{I}{y_{max}} = \frac{\frac{bd^3}{12}}{\frac{d}{2}} = \frac{bd^3}{12} \times \frac{2}{d} = \frac{bd^2}{6}$$

So, for a rectangular section,

$$z = \frac{bd^2}{6}$$

The basic stress equation

From the Engineer's Bending Equation (discussed above):

$$\frac{\sigma}{y} = \frac{M}{I}$$

Therefore:

$$\sigma = \frac{My}{I}$$

But:

$$z = \frac{I}{y_{max}}$$

Therefore:

$$\sigma = \frac{M}{z}$$

Or, rearranging:

$$z = \frac{M}{\sigma}$$

When I teach this material to students I express the opinion that the above equation is not immediately interesting or exciting and the reaction I get could be described as passive agreement. However, the above equation – unexciting as it may appear – forms the basis of all structural design. Let me explain.

Bending stress

If a bending moment (M) can be calculated – which it generally can if the loading and span of the beam are known (see Chapter 16) – and the permissible stress (σ) of the material is known (it can be obtained from science data tables), the required section size (z) can be determined. Once the required z value is known, a suitable timber beam size or 'off the peg' steel I section size can be selected, either by calculation or from tables, as shown in the following two examples.

Example 19.1: Timber beam

A timber beam spans 3.0 metres and carries a uniformly distributed load of 3.35 kN per metre run, as shown in Fig. 19.5. Headroom considerations dictate that a 225 mm deep timber section is used. If the allowable bending stress in timber is 6 N/mm², determine a suitable size (breadth × depth) for the beam.

Maximum bending moment (M) = $\dfrac{wL^2}{8}$ = $\dfrac{3.35 \times 3^2}{8}$ = 3.77 kN.m

$\sigma = \dfrac{M}{z}$ so, rearranging: $z = \dfrac{M}{\sigma}$

but $z = \dfrac{bd^2}{6}$ for a rectangular section

Therefore $\dfrac{bd^2}{6} = \dfrac{M}{\sigma}$

Rearranging: $b = \dfrac{6M}{\sigma d^2} = \dfrac{6 \times 3.77 \times 10^6 \text{ N.mm}}{6 \text{ N/mm}^2 \times 225^2 \text{ mm}^2}$

So minimum b = 74.5 mm
Therefore use a 75 mm wide × 225 mm deep timber beam.

(a) Elevation (b) Cross section

Fig. 19.5 Sizing a timber beam (Example 19.1).

Example 19.2: Steel beam design

A steel beam is to span 5 metres and will carry a load of 25 kN/metre, including its own weight, as shown in Fig. 19.6. If the permissible stress in the steel is 180 N/mm², select a suitable steel beam section from the tables.

From the information above, $w = 25$ kN/m and $L = 5$ m.

$$\text{Maximum bending moment } (M) = \frac{wL^2}{8} = \frac{25 \times 5^2}{8} = 78.1 \text{ kN.m}$$

$$z_{required} = \frac{M}{\sigma} = \frac{78.1 \times 10^6 \text{ N.mm}}{180 \text{ N/mm}^2} = 433\,889 \text{ mm}^3 = 433.9 \text{ cm}^3$$

Tables of the properties of standard steel beams should now be used. We need to select one that has a section modulus value of 433.9 cm³ or greater. The terminology used in the labelling of steel beams is explained in Chapter 24.

Possibilities include a 305 × 127UB37 steel beam ($z = 471$ cm³) and a 254 × 146UB37 steel beam ($z = 434$ cm³).

If the first of these is selected:

$$\text{Actual bending stress } (\sigma) = \frac{M}{z} = \frac{78.1 \times 10^6}{471\,000 \text{ mm}^3} = 165.8 \text{ N/mm}^2$$

As this is less than the permissible stress of 180 N/mm², this choice is fine. (Note: see Chapter 16 for the origin of $M = wL^2/8$.)

Repeat the above example with a span of 6 metres. You will find that the section modulus (z) value required this time is 625 000 mm³ and therefore a different steel beam section needs to be selected from the tables.

Fig. 19.6 Sizing a steel beam (Example 19.2).

Calculation of second moment of area (*I*) for symmetrical sections

As mentioned above, the *I* value for a rectangular section of breadth b and depth d is $bd^3/12$. Another useful piece of information is that the *I* value for a circle of diameter D is $\pi D^4/64$. Armed with the above information, it is straightforward to calculate *I* values for I sections or hollow rectangular sections (as illustrated in Fig. 19.7) or hollow circular sections. In each case, the shape can be considered as being the difference of the *I* values of two or more rectangles (as shown in Fig. 19.7) or the difference of the *I* values of two circles.

Consider the two examples shown in Fig. 19.8. In the first case, the *I* value for the I section can be determined by difference of *I* values for rectangular sections. In the second case, the *I* value for the hollow pipe is obtained by subtracting the *I* value for the inner circle from the *I* value for the outer circle. The calculations are given below.

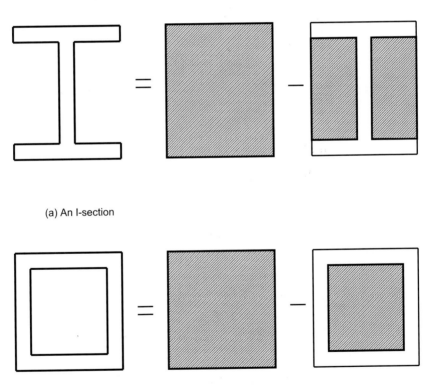

(a) An I-section

(b) A hollow rectangular section

Fig. 19.7 Calculation of second moment of area for common symmetrical shapes.

(a) An I-section (b) A hollow pipe

Fig. 19.8 Calculate the second moment of area (*I*) value for the above symmetrical shapes.

For the I section shown in Fig. 19.8 (a):

$$I = \frac{BD^3}{12} - \frac{bd^3}{12} = \frac{150 \times 300^3}{12} - \frac{130 \times 260^3}{12} = (337.5 \times 10^6) - (190.4 \times 10^6)$$

$$= 147.1 \times 10^6 \text{ mm}^4$$

For the hollow pipe section shown in Fig. 19.8 (b):

$$I = \frac{\pi D^4}{64} - \frac{\pi d^4}{64} = \frac{\pi}{64}(D^4 - d^4) = \frac{\pi}{64}(300^4 - 270^4) = 137 \times 10^6 \text{ mm}^4$$

Calculation of second moment of area (*I*) for unsymmetrical sections

The bad news is that, for unsymmetrical sections, determination of the second moment of area (*I*) value is a whole lot trickier. In brief, the procedure for unsymmetrical sections is as follows:

(1) Determine the position of the centroid of the section, using the approach outlined below. As you know from earlier in this chapter, the neutral axis always passes through the centroid of a section (assuming the section is made of the same material throughout). So, when you've determined the centroid position, you have also determined the level of the neutral axis.

(2) Once you know the neutral axis position, use the Parallel Axis Theorem (outlined below) to calculate the second moment of area (*I*) value.

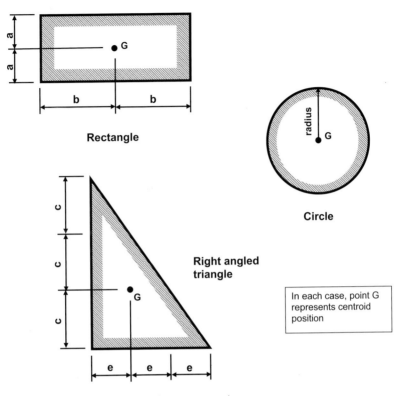

Fig. 19.9 Centroid positions for common shapes.

Centroids and how to locate them

The centroid is the geometric centre of area of a body, shape or section. If a body is of uniform density, the centre of gravity will be at the centroid. If a structural element is homogeneous (i.e. of the same material throughout) and experiences pure bending, the neutral axis (i.e. axis of zero stress) will pass through the centroid. Therefore the location of the centroid of a cross-section enables us to locate the level of the neutral axis (or neutral plane) relating to that cross-section.

Figure 19.9 shows the centroid positions of some common shapes. As we can see, the centroids of rectangles and circles occur at the centre of area (i.e. the obvious point), whereas the centroid of a right-angled triangle occurs one-third of the way along each side from the right-angled corner – or two-thirds of the way along from a 'pointed corner'.

Centroids of irregular shapes

An irregular shape, and the location of its centroid, is indicated in Fig. 19.10, from which it can be shown that:

$$A.\bar{x} = \sum(x.\delta A)$$

Fig. 19.10 Centroids of irregular shapes.

or:

$$\bar{x} = \sum \left(\frac{x.\delta A}{A} \right)$$

Similarly:

$$\bar{y} = \sum \left(\frac{y.\delta A}{A} \right)$$

where $\bar{x}$ and $\bar{y}$ are the distances from the y-axis and x-axis (respectively) to the centroid G. Note that the symbol Σ means 'sum of'. In other words, the dimension to the centroid of the total area from the appropriate axis or base line is equal to the sum of the area–distance products divided by total area.

Don't worry too much if you don't fully understand the mathematics above – it's the result, and its application, that is important.

Centroids of cross-sections which can be broken down into regular shapes

Most cross-sections encountered in civil engineering can be divided into constituent rectangles and triangles. The centroid positions in such cross-sections may be calculated using the above formulas.

Example

The beam shown in Fig. 19.11 can be divided into four rectangles as shown. The position of the section's centroid can be located from the following equations:

$$\bar{x} = \frac{A_1 x_1 + A_2 x_2 + A_3 x_3 + A_4 x_4}{A_1 + A_2 + A_3 + A_4}$$

$$\bar{y} = \frac{A_1 y_1 + A_2 y_2 + A_3 y_3 + A_4 y_4}{A_1 + A_2 + A_3 + A_4}$$

where:

A_1 = area of zone 1,
A_2 = area of zone 2, etc.

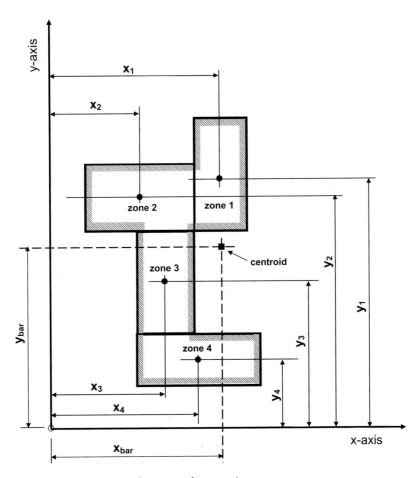

Fig. 19.11 Centroids of groups of rectangles.

x_1 = distance from y-axis to centroid of zone 1
x_2 = distance from y-axis to centroid of zone 2, etc.
y_1 = distance from x-axis to centroid of zone 1
y_2 = distance from x-axis to centroid of zone 2, etc.

Parallel Axis Theorem

The Parallel Axis Theorem can be used to calculate I values (i.e. second moment of area values) for sections that can be divided into individual rectangular parts. (For a rectangular section, $I = bd^3/12$.)

First, the neutral axis level (i.e. centroid position) has to be determined, in the manner previously discussed. Consider the rectangular element emphasised in Fig. 19.12, which forms part of a larger cross-section. It can be shown that:

$$I_{XX} = I_{CC} + Ah^2$$

or, for a rectangle,

$$I_{XX} = (bd^3/12) + bdh^2$$

where:

I_{XX} = second moment of area of the rectangular element about the neutral axis of the composite section (i.e. about axis X–X)
I_{CC} = second moment of area of the rectangular element about the axis through its centroid (i.e. about axis C–C)
A = area of rectangular element

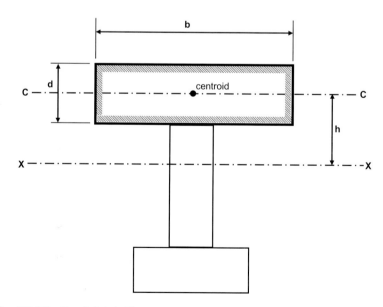

Fig. 19.12 Parallel Axis Theorem.

b = breadth of rectangular element
d = depth of rectangular element
h = distance between centroidal axis of the rectangular element and the centroidal axis of the composite section.

The total I_{xx} for the composite section is equal to the sum of the I_{xx} terms for the individual parts.

Example 19.3: Bending stresses in a T section

A beam with the cross-section shown in Fig. 19.13 is simply supported and carries a maximum bending moment of 16.0 kN.m. Calculate:

- The neutral axis position.
- Maximum tensile stress.
- Maximum compressive stress.

Fig. 19.13 Calculation of second moment of area for a T section (Example 19.3). All dimensions are in mm.

Table 19.1 Calculation of second moment of area for Example 19.3

(1)	(2)	(3)	(4)	(5)	(6)	(7)	(8)	(9)
Zone	b (mm)	d (mm)	A (mm²)	y (mm)	Ay (mm³)	h (mm)	Ah² (mm⁴) (x10⁶)	I = bd³/12 (mm⁴) (x10⁶)
1	100	20	2000	10	20 000	42.4	3.59	0.07
2	20	130	2600	85	221 000	32.6	2.76	3.66
Sum			4600		241 000		6.35	3.73

In column (7): 42.4 = 52.4 − 10
32.6 = 85 − 52.4

Solution

Use the top edge of the beam as a datum from which to calculate distances. I suggest a methodical way to approach this problem would be to do the calculations in tabular form − see Table 19.1.

Split the cross-section into two rectangles: let the 'cross-bar' be zone 1 and the stem of the T be zone 2. The breadth (b) and depth (d) of each zone are given in columns (2) and (3) of Table 19.1. In each case, these are multiplied together to give the area of each zone, shown in column (4).

The y values are the vertical distances from the top of the section (the datum level) to the centroids of each zone. It can be seen from Fig. 19.13 that these values are 10 mm (half of 20 mm) for zone 1 and 85 mm (20 mm + half of 130 mm) for zone 2. These values are given in column (5) of Table 19.1.

The values given in column (6) are A (from column (4)) multiplied by y (from column (5)). From column (6) it can be seen that the sum of the Ay values is 241,000 mm³ and from column (4) the sum of the A values (i.e. the total area of the section) is 4600 mm². So the distance to the section's centroid from the top (y_{bar}) is calculated as follows:

$$y_{bar} = \frac{\Sigma(Ay)}{\Sigma A} = \frac{241{,}000 \text{ mm}^3}{4600 \text{ mm}^2} = 52.4 \text{ mm}$$

(This is the answer to part 1 of the question.)

Now that the position of the centroid of the section has been determined, the distances, h, from the section's centroidal axis to the centroids of the individual zones (depicted as h_1 and h_2 in Fig. 19.13) can be calculated. These figures are given in column (7) of Table 19.1.

Column (8), Ah^2, is A (from column (4)) multiplied by h^2 (from column (7)). Column (9) is the I value (= $bd^3/12$) for each rectangular zone.

When discussing the Parallel Axis Theorem above, we saw that:

$$I_{xx} = Ah^2 + \frac{bd^3}{12}$$

So, I_{xx} is the sum of all the figures in column (8) and all the figures in column (9):

$$I_{xx} = (6.35 + 3.73) \times 10^6 = 10.08 \times 10^6 \text{ mm}^4$$

Now we know the I value we can calculate the bending stresses. Earlier in this chapter we saw that:

$$\sigma = \frac{My}{I}$$

where:

σ = bending stress
M = bending moment
y = distance from neutral axis to top or bottom of section
I = second moment of area.

In this example:

$M = 16.0$ kN.m $= 16.0 \times 10^6$ N.mm (given in question)

$I = 10.08 \times 10^6$ mm^4 (calculated above)

$y = 52.4$ mm (to top of section)

$y = (150-52.4) = 97.6$ mm (to bottom of section)

As this beam is simply supported, the maximum tensile stress occurs in the bottom of the section and the maximum compressive stress occurs in the top. So:

Max tensile stress (bottom of section) =

$$\frac{My}{I} = \frac{16.0 \times 10^6 \text{ N.mm} \times 97.6 \text{ mm}}{10.08 \times 10^6 \text{ mm}^4} = 154.9 \text{ N/mm}^2$$

Max compressive stress (top of section) =

$$\frac{My}{I} = \frac{16.0 \times 10^6 \text{ N.mm} \times 52.4 \text{ mm}}{10.08 \times 10^6 \text{ mm}^4} = 83.2 \text{ N/mm}^2$$

Example 19.4: Bending stresses in a non-symmetrical I section

Determine the maximum bending moment that can be applied to the simply supported beam whose cross-section is shown in Fig. 19.14 if:

- Maximum tensile stress = 2.0 N/mm².
- Maximum compressive stress = 20 N/mm².

Fig. 19.14 Calculation of second moment of area for a non-symmetrical I section (Example 19.4). All dimensions are in mm.

As in the previous example, we will use a table (Table 19.2) to calculate the neutral axis position and the second moment of area (I) value.

Table 19.2 Calculation of second moment of area for Example 19.4

(1)	(2)	(3)	(4)	(5)	(6)	(7)	(8)	(9)
Zone	b (mm)	d (mm)	A (mm^2)	y (mm)	Ay (mm^3)	h (mm)	Ah^2 (mm^4) (x 10^6)	I = bd^3/12 (mm^4) (x 10^6)
1	320	30	9 600	15	144 000	372.7	1 333.50	0.72
2	30	600	18 000	330	5 940 000	57.7	59.93	540.00
3	440	40	17 600	650	11 440 000	262.3	1 210.90	2.35
Sum			45 200		17 524 000		2 604.33	**543.07**

In column (7): 372.7 = 387.7 −15
57.7 = 387.7 − 330
262.3 = 650 − 387.7

Again, we'll use the top edge of the beam as the datum. (Note: it doesn't matter what level you use as your datum, provided you are consistent throughout.)

From Table 19.2:

$$\bar{y} = \frac{\Sigma(Ay)}{\Sigma A} = \frac{17\,524\,000}{45\,200} = 387.7 \text{ mm from top (342.3 mm from bottom)}$$

$$I_{xx} = (2604.33 + 543.07) \times 10^6 = 3147.4 \times 10^6 \text{ mm}^4$$

Unlike the previous example, it is not bending stresses we need to calculate this time. We need to determine the bending moments associated with particular values of tensile and compressive stress.

From the Engineers' Bending Equation:

$$\frac{\sigma}{y} = \frac{M}{I}$$

Therefore, rearranging:

$$M = \frac{\sigma y}{I}$$

Using the above equation we can calculate the moment that would cause the maximum (compressive) stress in the top of the beam and the moment that would cause the maximum (tensile) stress in the bottom:

In top of beam:

$$M = \frac{\sigma_{comp} \times I}{y_{top}} = \frac{20 \text{ N/mm}^2 \times 3147.4 \times 10^6 \text{ mm}^4}{387.7 \text{ mm}} = 162.4 \times 10^6 \text{ N.mm} = 162.4 \text{ kN.m}$$

In bottom of beam:

$$M = \frac{\sigma_{comp} \times I}{y_{btm}} = \frac{2.0 \text{ N/mm}^2 \times 3147.4 \times 10^6 \text{ mm}^4}{342.3 \text{ mm}} = 18.4 \times 10^6 \text{ N.mm} = 18.4 \text{ kN.m}$$

So the maximum bending moment that could be applied to the beam would be the lesser of the two figures calculated above, i.e. 18.4 kN.m.

What you should remember from this chapter

- A simply supported beam subjected to bending (in a sagging mode) will experience maximum tensile stress in the bottom and maximum compressive stress in the top.
- The magnitude of the stress varies linearly between the top of the section and the bottom.

- The level at which there is no stress is called the neutral axis. For symmetrical sections of the same material throughout, the neutral axis occurs half way down the section.
- Before a given cross-section can be analysed for stress, the second moment of area needs to be calculated. While this is relatively straightforward for symmetrical sections, it is more complicated for non-symmetrical sections, for which the Parallel Axis Theorem must be used.

Tutorial questions

(1) A timber beam of rectangular cross-section 75 mm wide and 300 mm deep carries a 5 kN point load at the mid point of a simply supported span of 4 metres. Determine the maximum bending stress in the beam.

(2) A steel beam with a symmetrical I-shaped cross-section sustains a uniformly distributed load of 25 kN/m over a simply supported 3 metre span. The cross-section dimensions (all in millimetres) are given in Fig. 19.15. Calculate:
   (a) The maximum bending stress in the beam.
   (b) The radius of curvature of the beam, given $E = 205$ kN/mm².
   (c) The bending stress at the top of the web in the beam at the location of maximum bending moment.

(3) A hollow tube of 50 mm external diameter and 44 mm internal diameter is subjected to a bending moment of 0.50 kN.m. Determine the maximum bending stress.

Fig. 19.15 Tutorial question no. 2.

Fig. 19.16 Tutorial question no. 4.

(4) The reinforced concrete hollow rectangular section shown in Fig. 19.16 comprises the cross-section of a 5 metre long beam, which sustains a 15 kN/m uniformly distributed load. Calculate:
  (a) The maximum bending stress in the beam.
  (b) The radius of curvature of the beam given $E = 20$ kN/mm².
(5) Figure 19.17 shows the cross-sectional geometry of a steel beam. The cross-section of the beam is symmetrical about both the X–X and Y–Y

Fig. 19.17 Tutorial question no. 5.

Fig. 19.18 Tutorial question no. 6.

axes. The beam spans 4 metres and supports a uniformly distributed load of 4 kN/m. Calculate the following:
(a) The second moment of area about the X–X axis (I_{xx}).
(b) The maximum bending moment in the beam.
(c) The maximum bending stress.
(d) The strain corresponding to the stress calculated in (c) if Young's Modulus (E) for the steel beam is 205 kN/mm².

(6) Figure 19.18 shows the geometry of a steel T section. A beam is constructed from this section and required to sustain a maximum bending moment of 75 kN.m. Calculate:
(a) The depth of the centroidal (X–X) axis from the top of the section.
(b) The second moment of area about the X–X axis (I_{xx}).
(c) The maximum bending stress in the beam when it is subjected to the maximum bending moment of 75 kN.m.

(7) Figure 19.19 shows the cross-section of a steel beam. The section is symmetrical about the Y–Y axis and the X–X axis passes through the centroid of the section and forms the neutral axis. Calculate:
(a) The depth of the centroidal (X–X) axis from the top of the section.
(b) The second moment of area about the X–X axis (I_{xx}).
(c) The maximum bending stress in the beam when it is subjected to a maximum bending moment of 50 kN.m.

Fig. 19.19 Tutorial question no. 7.

Tutorial answers

(1) 4.44 N/mm².
(2) (a) 74.6 N/mm²;(b) 274.7 metres; (c) 59.7 N/mm².
(3) 101.8 N/mm².
(4) (a) 23.3 N/mm²; (b) 150.5 metres.
(5) (a) 38.9×10^6 mm⁴; (b) 8 kN.m; (c) 25.7 N/mm²; (d) 1.25×10^{-4}.
(6) (a) 97.5 mm; (b) 89.2×10^6 mm⁴; (c) 170.3 N/mm².
(7) (a) 85 mm; (b) 16.35×10^6 mm⁴; (c) 290.6 N/mm².

Combined bending and axial stress

Introduction

In Chapter 18 we studied direct stresses. We found that the value of direct stress is constant across a cross-section and is equal to the axial force (P) divided by the cross-sectional area (A). In Chapter 19 we investigated bending stresses. There we found that the value of bending stress is not constant across a cross-section (in fact, it varies linearly) and that its maximum value is given by the bending moment (M) divided by its section modulus (z).

In this chapter we will see what happens when direct stresses and bending stresses are combined.

Combined stresses by formula

$$\text{Direct (axial) stress } (\sigma) = \frac{P}{A} \text{ (from Chapter 18)}$$

$$\text{Maximum bending stress } (\sigma) = \frac{M}{z} \text{ (from Chapter 19)}$$

These two equations can be combined, as shown below.

$$\text{Combined bending and axial stress} = \frac{P}{A} \pm \frac{M}{z}$$

It is not easy to see how this equation may be applied. To assist in this, look at Fig. 20.1. The two diagrams show the elevation of a column before

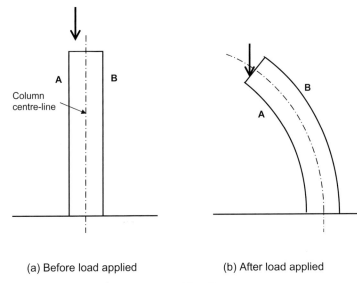

(a) Before load applied (b) After load applied

Fig. 20.1 Column with eccentric axial loading.

and after an eccentric longitudinal load is applied. As you see, the left-hand side of the column (side A) is pushed down from its original position under the effects of the axial load, while the right-hand side of the column (side B) is pulled up. This suggests that side A is experiencing compression while side B is undergoing tension.

Each of the nine diagrams shown in Fig. 20.2 represents a plan view of a column which is square in cross-section. The four sides are labelled A, B, C and D. In each diagram, the large black blob represents the position at which the longitudinal load is applied.

In each case, determine which side(s) of the column experience tension and which side(s) experience compression. To make things simpler, we will introduce a +/− sign convention as follows:

- If a side is pushed downwards under the applied load, it experiences compression (+).
- If a side is pulled upwards under the applied load, it experiences tension (−).

The answers, in the form of + and − signs, are shown in Fig. 20.3.

Keep the above exercise in mind as you progress through this chapter. It will help you to determine whether a + or a − sign is required at various points in your calculations.

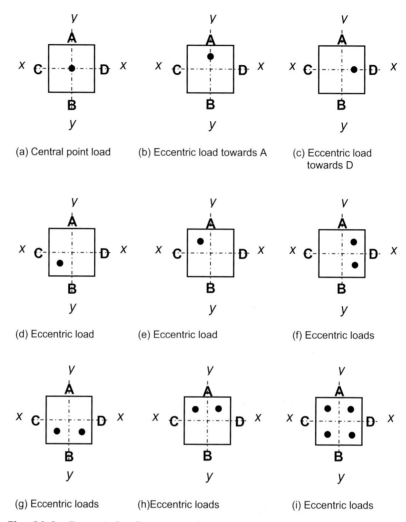

Fig. 20.2 Eccentric loading on a column.

Another way of looking at combined bending and axial stress

Consider the rectangular cross-section shown in Fig. 20.4 (a). In Chapter 18 we learned that direct (or axial) stress has a value P/A which is constant across the cross-section. This is illustrated in Fig. 20.4 (b). By contrast, we learned in Chapter 19 that the value of bending stress varies linearly across a cross-section, with a maximum value of M/z. This was illustrated in Fig. 19.3 and is shown again, here, in Fig. 20.4 (c). If we combine the two graphs,

Combined bending and axial stress 217

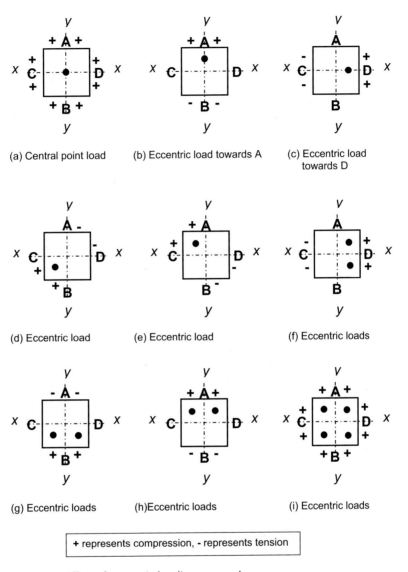

Fig. 20.3 Effect of eccentric loading on a column.

the result depends on the relative values of P/A and M/z. If P/A is greater than M/z, the combined graph will appear as in Fig. 20.4 (d). But if P/A is less than M/z, the combination is shown in Fig. 20.4 (e). Note that this last case gives rise to tensile stresses when it is negative.

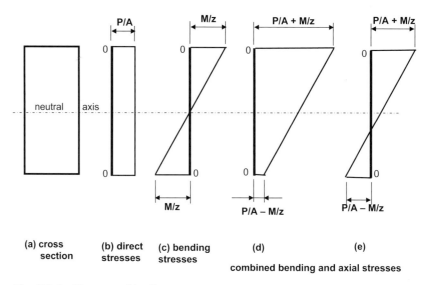

Fig. 20.4 Stress combinations.

The formulas

I have said earlier in the book that I'm not a great fan of 'magic formulas' into which students can plug numbers and produce a (possibly incorrect) answer without a great deal of understanding of what they're doing. However, calculations for combined bending and axial stress situations are dependent on certain formulas – but you have to know when to use a plus sign and when to use a minus sign. And, as with any formula, you have to understand what the various terms mean.

Earlier in this chapter we encountered the following equation:

$$\text{Combined bending and axial stress} = \frac{P}{A} \pm \frac{M}{z}$$

Now, a force P acting at an eccentricity e from the centre line of a cross-section will apply a moment of $(P \times e)$ at that centre line. So:

$M = Pe$

Also, in Chapter 19 we learned that $z = I/y$. From this we can generate two further equations for combined bending and axial stress, as follows:

$$\text{Combined bending and axial stress} = \frac{P}{A} \pm \frac{My}{I}$$

or:

$$\text{Combined bending and axial stress} = \frac{P}{A} \pm \frac{Pey}{I}$$

For a reminder of what all the symbols mean, see Fig. 20.5.

Combined bending and axial stress

Area $A = bd$

$M = Pe$

$z = bd^2/6$ (see Chapter 19)

$I = bd^3/12$ (see Chapter 19)

y = vertical distance from neutral axis to point of interest

Fig. 20.5 Symbols in combined bending and axial stress equation.

Example 20.1

A force of 200 kN acts vertically downwards on a column of cross-sectional dimensions 400 mm × 300 mm. The force acts at an eccentricity of 100 mm along the Y–Y axis from the centre of the section, as shown in Fig. 20.6 (a). Calculate the stress in the column at the following positions:

- Along the 'top' face of the column (position A).
- At the point of application of the load (point K).
- At the centroid of the cross-section (point L).
- At a point 50 millimetres 'below' the centre line (point M).
- Along the 'bottom' face of the column (position B).

We know the following:

$P = 200$ kN (or 200×10^3 N)

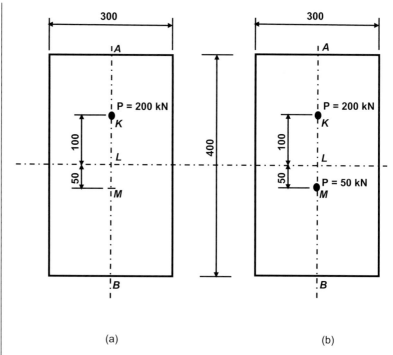

Fig. 20.6 Worked example 20.1.

$A = bd = (300 \text{ mm} \times 400 \text{ mm}) = 120{,}000 \text{ mm}^2$

$e = 100 \text{mm}$

$M = Pe = (200 \times 10^3 \text{ N} \times 100 \text{ mm}) = 20 \times 10^6 \text{ N.mm}$

$I = \dfrac{bd^3}{12} = \dfrac{300 \times 400^3}{12} = 1.6 \times 10^9 \text{ mm}^4$

y is the distance from the centroidal axis (X–X) to the position at which we're interested in calculating the stress. Its values for positions A, K, L, M and B are respectively 200, 100, 0, 50 and 200 millimetres.

Signs are also important. As the force P is pushing down on the upper part of the section, it will induce compression (+) for points A and K, zero for L, and tension (–) for points M and B.

$$\sigma = \dfrac{P}{A} \pm \dfrac{My}{I}$$

For point A: $\sigma_A = \dfrac{200 \times 10^3}{120{,}000} + \dfrac{20 \times 10^6 \times 200}{1.6 \times 10^9} = 1.67 + 2.5 = 4.17 \text{ N/mm}^2$

For point K: $\sigma_K = \dfrac{200 \times 10^3}{120{,}000} + \dfrac{20 \times 10^6 \times 100}{1.6 \times 10^9} = 1.67 + 1.25 = 2.92 \text{ N/mm}^2$

For point L: $\sigma_L = \dfrac{200 \times 10^3}{120{,}000} + \dfrac{20 \times 10^6 \times 0}{1.6 \times 10^9} = 1.67 + 0 = 1.67 \text{ N/mm}^2$

For point M: $\sigma_M = \dfrac{200 \times 10^3}{120{,}000} - \dfrac{20 \times 10^6 \times 50}{1.6 \times 10^9} = 1.67 - 0.625 = 1.045 \text{ N/mm}^2$

For point B: $\sigma_B = \dfrac{200 \times 10^3}{120{,}000} - \dfrac{20 \times 10^6 \times 200}{1.6 \times 10^9} = 1.67 - 2.5 = -0.83 \text{ N/mm}^2$

Now let's make the problem slightly harder. Let's suppose that, in addition to the 200 kN force shown above, a 100 kN force acts at point M, as illustrated in Fig. 20.6 (b).

The overall moment about the X–X axis is now:

$M = (200 \times 10^3 \text{ N} \times 100 \text{ mm}) - (100 \times 10^3 \text{ N} \times 50 \text{ mm}) = 15 \times 10^6 \text{ N.mm}$

The total force, P, is now:

$(200 \text{ kN} + 100 \text{ kN}) = 300 \text{ kN}$ (or 300×10^3 N)

So, the first term of the equation is now:

$$\dfrac{P}{A} = \dfrac{300 \times 10^3}{120{,}000} = 2.5 \text{ N/mm}^2$$

The other quantities remain the same. So now the stresses are as follows:

For point A: $\sigma_A = 2.5 + \dfrac{15 \times 10^6 \times 200}{1.6 \times 10^9} = 2.5 + 1.875 = 4.375 \text{ N/mm}^2$

For point K: $\sigma_K = 2.5 + \dfrac{15 \times 10^6 \times 100}{1.6 \times 10^9} = 2.5 + 0.938 = 3.438 \text{ N/mm}^2$

For point L: $\sigma_L = 2.5 + \dfrac{15 \times 10^6 \times 0}{1.6 \times 10^9} = 2.5 + 0 = 2.5 \text{ N/mm}^2$

For point M: $\sigma_M = 2.5 - \dfrac{15 \times 10^6 \times 50}{1.6 \times 10^9} = 2.5 - 0.047 = 2.453 \text{ N/mm}^2$

For point B: $\sigma_B = 2.5 - \dfrac{15 \times 10^6 \times 200}{1.6 \times 10^9} = 2.5 - 1.875 = +0.625 \text{ N/mm}^2$

These stresses, for each of the two cases considered, are tabulated in Table 20.1.

Table 20.1	Stresses derived from Example 20.1		
Point	Description of point	200 kN load only	200 kN load + 100 kN load
A	'Top' face of column	+4.17	+4.375
K	100 mm above centre	+2.92	+3.438
L	At centre of column section	+1.67	+2.5
M	50 mm below centre	+1.045	+2.453
B	'Bottom' face of column	−0.83	+0.625

The figures in the right-hand columns are stresses, expressed in N/mm² units

Beware the difference between e and y

Many students are puzzled as to the distinction between e and y. This distinction is crucial to the understanding of problems involving combined bending and axial stress, and is as follows:

- e represents the eccentricity of the load(s) – that is, the distance from the point of action of the load to the relevant centroidal axis (axis X–X in the above case). In the above example, e = 100 mm for the 200 kN load and 50 mm for the 100kN load.
- y represents the distance from the centroidal axis to the point at which we wish to know the stress.

Maximum and minimum values of stress

Examine the figures in Table 20.1. You will see that in each case the maximum stress occurs in the 'top' face (position A) and the minimum stress occurs in the 'bottom' face (position B).

The values of maximum and minimum stresses are particularly important to engineers, as we design a column (or other structural element) to sustain the worst stress to which it is likely to be subjected. This 'worst' stress is usually the maximum value, but the minimum value is of interest too, especially if it is negative (as it was in the case of point B in Example 20.1 above). A negative value of stress suggests that tension is being experienced and in many situations we need to avoid tensile stresses. More of that later.

Combined stresses in two dimensions

So far we have considered stresses in one dimension only. (For example, in Fig. 20.6, points A, K, L, M and B all lie on the same vertical line.) This is fine for situations where the loads happen to act directly on centroidal axes, but what happens if they don't?

Combined bending and axial stress   223

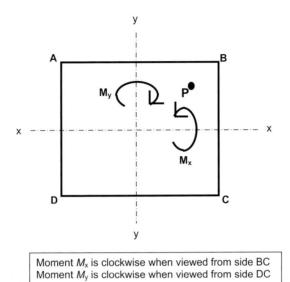

Moment M_x is clockwise when viewed from side BC
Moment M_y is clockwise when viewed from side DC

Fig. 20.7 Stresses caused by rotation about both axes – general case.

Examine Fig. 20.7, which shows a column cross-section on which a load P is acting. P acts at a point which is eccentric from the column's centroid in both directions – in other words, the point is on neither the X–X or Y–Y axes. The four corners of the column are labelled A, B, C and D.

The eccentric load P will induce a moment about each of the axes X–X and Y–Y. We will call these moments M_x and M_y respectively.

$z_x = bd^2/6$ and $z_y = db^2/6$

(z, the section modulus, was introduced in Chapter 19.)

The stresses at the four corners (A, B, C and D) of the column can be calculated from the following equations:

$$\sigma_A = \frac{P}{A} + \frac{M_x}{z_x} - \frac{M_y}{z_y}$$

$$\sigma_B = \frac{P}{A} + \frac{M_x}{z_x} + \frac{M_y}{z_y}$$

$$\sigma_C = \frac{P}{A} - \frac{M_x}{z_x} + \frac{M_y}{z_y}$$

$$\sigma_D = \frac{P}{A} - \frac{M_x}{z_x} - \frac{M_y}{z_y}$$

Example 20.2

As in Example 20.1 above, a column of cross-sectional dimensions 400 mm × 300 mm experiences a load of 200 kN. This time though, the load is applied eccentrically to both axes, as shown in Fig. 20.8. Calculate the stress at each of the four corners of the column (A, B, C and D).

$P = 200$ kN (or 200×10^3 N)

$A = (300 \text{ mm} \times 400 \text{ mm}) = 120{,}000 \text{ mm}^2$

$M_x = +(200 \times 10^3 \text{ N} \times 100 \text{ mm}) = 20 \times 10^6$ N.mm

$M_y = +(200 \times 10^3 \text{ N} \times 50 \text{ mm}) = 10 \times 10^6$ N.mm

$z_x = bd^2/6 = 300 \times 400^2/6 = 8.0 \times 10^6 \text{ mm}^3$

$z_y = db^2/6 = 400 \times 300^2/6 = 6.0 \times 10^6 \text{ mm}^3$

Note that M_x and M_y are both positive because they both act in the same direction as the general case shown in Fig. 20.7.

$$\sigma = \frac{P}{A} \pm \frac{M_x}{z_x} \pm \frac{M_y}{z_y}$$

$$\sigma = \frac{200 \times 10^3}{120{,}000} \pm \frac{20 \times 10^6}{8 \times 10^6} \pm \frac{10 \times 10^6}{6 \times 10^6} \text{ N/mm}^2$$

$\sigma = 1.67 \pm 2.5 \pm 1.67$ N/mm²

Fig. 20.8 Worked example 20.2.

So the stresses at the four corners are:

$\sigma_A = 1.67 + 2.5 - 1.67 = +2.5 \text{ N/mm}^2$

$\sigma_B = 1.67 + 2.5 + 1.67 = +5.84 \text{ N/mm}^2$

$\sigma_C = 1.67 - 2.5 + 1.67 = +0.84 \text{ N/mm}^2$

$\sigma_D = 1.67 - 2.5 - 1.67 = -2.5 \text{ N/mm}^2$

Note the negative value of stress at point D – it indicates that tensile stress is being experienced there.

Pressure on foundations

The principles outlined above regarding eccentric loads on columns are equally applicable to eccentric loads on foundations. Columns in buildings have to be supported at their base by a foundation, whose function is to safely transmit all the loads from a structure safely into the ground (see Chapter 1). A concrete pad (or isolated) footing is often used, as illustrated in Chapter 3.

In the design of pad foundations it is important to ensure that the permissible ground bearing pressure (that is, the maximum pressure that the ground can sustain) is not exceeded. It is therefore important to be able to calculate the actual pressure at any point in the foundation. In practice, the maximum or minimum pressures occur at one of the four corners, so it is sufficient to calculate the actual pressure at each of the corners.

Figure 20.7, which we referred to earlier when we were considering stresses in columns, is equally applicable to the general case for pressure on foundations. It is a plan view of a rectangular concrete pad foundation whose four corners are labelled A, B, C and D.

The two centroidal axes are labelled X–X and Y–Y. An eccentric load P acts at a position that causes a clockwise moment (as viewed from side BC) about axis X–X and a clockwise moment (as viewed from side DC) about axis Y–Y. The pressures at the corners A, B, C and D are given by the four equations discussed earlier.

(Note: Although we used N and mm units when calculating stresses in columns, the larger forces and dimensions in foundations suggest that kN and metres are more suitable units when calculating pressures in foundations.)

Example 20.3

Calculate the pressure at each corner of the foundation shown in Fig. 20.9. The 80 kN load will cause a clockwise rotation about the x-axis (as viewed from side BC) which is the same as that assumed in the general case of Fig. 20.8. Hence the positive sign in the M_x calculation below.

The 80 kN load will cause an anticlockwise rotation about the y-axis (as viewed from side DC), which is opposite in direction from the clockwise rotation assumed in the general case of Fig. 20.8. Hence the negative sign in the M_y calculation below.

$P = 80$ kN

$M_x = +(80 \text{ kN} \times 0.2 \text{ m}) = +16$ kN.m

$M_y = -(80 \text{ kN} \times 0.7 \text{ m}) = -56$ kN.m

$A = (3.0 \times 1.5) = 4.5$ m²

$$Z_x = \frac{bd^2}{6} = \frac{3 \times 1.5^2}{6} = 1.125 \text{ m}^3$$

$$Z_y = \frac{db^2}{6} = \frac{1.5 \times 3^2}{6} = -2.25 \text{ m}^3$$

$$\sigma = \frac{P}{A} \pm \frac{M_x}{Z_x} \pm \frac{M_y}{Z_y}$$

$$\sigma = \frac{80}{4.5} \pm \frac{16}{1.125} \pm \frac{-56}{2.25}$$

$\sigma = 17.78 \pm 14.22 \pm (-24.89)$

$\sigma_A = 17.78 + 14.22 - (-24.89) = +56.89$ kN/m²

$\sigma_B = 17.78 + 14.22 + (-24.89) = +7.11$ kN/m²

$\sigma_C = 17.78 - 14.22 + (-24.89) = -21.33$ kN/m²

$\sigma_D = 17.78 - 14.22 - (-24.89) = +28.45$ kN/m²

As the pressure at corners C is negative, this suggests that tension occurs at this point. In other words, the foundation would tend to lift off the ground at point C, which is obviously not desirable in practice!

Fig. 20.9 Example of eccentric foundation loading.

What you should remember from this chapter

This chapter explains how to combine axial and bending stresses in a column or a foundation. The calculation procedure has been outlined to obtain the overall stress (or pressure) at any point in a column (or foundation). Watch out for the signs (+ or −) and be aware that a negative stress indicates that tension is occurring at the point concerned.

Tutorial questions

Calculate the stresses at each of the four corners (A, B, C and D) of the four examples illustrated in Fig. 20.10. In each case, identify the points (if any) at which tension occurs. (Note: In each case, the loads act 'into the paper'.)

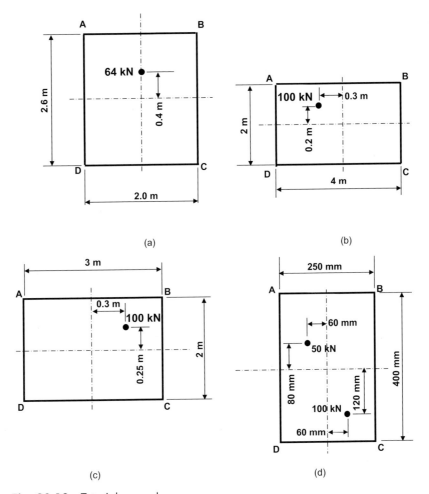

Fig. 20.10 Tutorial examples.

Tutorial answers

Values given are at points A, B, C and D respectively.
(a) +23.67, +23.67, +0.95, +0.95 kN/m².
(b) +25.63, +14.37, −0.63, +10.63 kN/m².
(c) +19.17, +39.17, +14.17, −5.83 kN/m².
(d) −419, +1019, +3419, +1981 kN/m².

Structural materials: concrete, steel, timber and masonry

Introduction

This book is primarily concerned with the basics of structural analysis. Up till now we haven't paid much attention to the material that a beam, column or slab might be made of. There are, of course, many materials available for us to use, but in this chapter we will confine our discussion to the four main structural materials, namely concrete, steel, timber and masonry.

Both architects and structural engineers need to decide at an early stage what material (or combination of materials) they are going to use in a particular project. But it's difficult to make such a decision if you don't know anything about the various materials. The purpose of this chapter is to discuss the different materials available to the construction professional.

Which is the best material?

A natural question at this stage is: which is the best structural material? Well, it depends on what you mean by 'best'. Does 'best' mean strongest, stiffest, cheapest, most readily available or most attractive? Or all of these? Or maybe none of these?

A moment's consideration would lead us to conclude that there is no one building material that is the best in all respects. If there were, then every building structure in the entire world would be built out of that one material. Clearly this isn't the case. If we look at the world around us, we see buildings made of brickwork or stonework, timber buildings, and buildings with frames of steel or reinforced concrete. In certain parts of the world we see buildings constructed of ice, mud or bamboo. It is apparent that there are many different materials that can be used in building, each of which has its advantages and disadvantages.

The kettle analogy

If you look at your everyday surroundings you will notice that particular objects tend to be made of certain materials. This is because these materials are particularly suitable for given applications. For example, car tyres are made of rubber, windows are made of glass, pens are usually plastic.

We also know that certain materials are patently unsuitable for certain applications. For example:

- Contact lenses are never made from steel.
- Aircraft fuselages are never constructed from brickwork.
- Computers are never made out of concrete.
- Radiators are never made from plastic (although perhaps they could be?).

Consider a kettle as an example. If you review the desired properties of a kettle, you might come up with some or all of the following:

- Strength: the kettle must be strong enough to contain water and to resist the pressure of steam building up inside it. It must also be strong enough not to break if dropped onto a hard floor surface.
- Thermal properties: the kettle must be able to resist the temperature of boiling water and must not break, melt or otherwise deform at such temperatures. It must also be able to cope with sudden changes of temperature, for example if cold water is poured into a recently boiled kettle.
- Rigidity: the kettle must not deform under water or steam pressure.
- Disposability: what will happen to the kettle at the end of its life?
- Availability of materials: the materials must be readily available in the quantities required for mass production of kettles.
- Manufacturing costs: the manufacturing process must be streamlined so that kettles are produced as cheaply as possible.
- Durability: the kettle should not readily rot, corrode or otherwise degrade in use.
- Waterproofness: the kettle shouldn't leak.
- Attractiveness: the kettle should be sufficiently good looking that people would want to buy it.

A manufacturer of kettles has to find a material that has all the above properties. Until the late 1970s, all kettles were made of steel; then plastics were developed that could cope with high temperatures without deforming. Nowadays most kettles are made of plastic because there are plastics available that meet the above requirements and are cheaper than steel. Let's consider the consequences of making kettles out of other materials:

- A timber kettle is possibly more expensive to manufacture. It would be difficult to achieve a waterproof seal and the timber would rot quickly in such a damp, steamy environment unless preservatives were used – which may be poisonous!

- It would be difficult (and therefore uneconomic) to create a concrete kettle to the required dimensions; otherwise it would be too heavy. Also, the surface of the concrete might tend to flake off or dissolve inside into the water being boiled.
- A masonry kettle would be impractical for the same reasons as a concrete one, with formation of waterproof joints being an additional problem.

So what was the purpose of this diversion into the preferred properties of a kettle? Well, some of the properties listed above, desirable in the manufacture of kettles, are also important properties of the materials to be used in structures. Let's examine some of these desirable properties in more detail.

Factors to be considered in material selection

Availability

Construction materials are used in large quantities and therefore need to be readily available. Stone and clay are extracted in most parts of the United Kingdom, hence masonry (stonework, brickwork and blockwork) is widely used in domestic construction. (For example, until the 1960s every building in the Scottish city of Aberdeen was built out of granite, which was readily available locally from one massive quarry.) In some parts of the world, other locally available materials are excellent for construction. Also, the local labour force is likely to be familiar with the use of locally available materials.

Strength

Materials need to be strong enough (in tension and/or compression) for their intended purpose. Clearly, some materials are stronger than others. Selection of too weak a material for a particular application is an obvious mistake, but selection of a needlessly strong material is also undesirable.

Stiffness

Stiffness, or rigidity, is not to be confused with strength: some strong materials are not stiff (e.g. rope) and some stiff materials are not particularly strong (e.g. glass). The stiffer a material, the less it will deflect. The stiffness of a material is proportional to its Young's Modulus value. (For the derivation of Young's Modulus, see Chapter 18.) Typical Young's Modulus values for the materials being considered in this chapter are:

- Steel: 210 kN/mm^2
- Aluminium: 71 kN/mm^2
- Concrete: 14 kN/mm^2
- Timber: 5–10 kN/mm^2

It can be seen from the above that steel is by far the stiffest of the common structural materials – for a given cross-section steel is three times as stiff as aluminium, 15 times as stiff as concrete and over 20 times as stiff as timber. But remember, this is for a constant cross-section, so these relative stiffnesses will vary according to the cross-section used.

We saw in Chapter 1 that deflection needs to be controlled, but it is less critical in some applications than others. A super-stiff material, therefore, is not always required or even desirable.

Speed of erection

Some building types can be erected more quickly than others. For example, a steel-framed structure can be completed far more quickly than a masonry one. But speed of construction is not always critical and there may well be a trade-off between speed and cost. Being told that a building could be built twice as fast for twice the cost greatly concentrates the mind!

Cost/economics

A complex issue. Architects and engineers are always looking to minimise cost. There is an old saying that an engineer can do for a penny what anyone can do for two pence. We have to consider the cost of the raw materials, the cost of conversion of the material into its usable form, transportation costs and associated labour costs.

Ability to accommodate movement

All buildings tend to move. Some materials can accommodate this better than others. For example, brickwork can cope with movement more readily than a steel-framed structure can.

Durability

Some materials rot, decompose, corrode or spall, etc. over time. Some materials do this more readily than others; in other words, some materials are less durable than others. Maintenance costs and programmes need to be taken into account. For example, it is well-known that the Forth Rail Bridge in Scotland is repainted on a three- to five-year cycle to control corrosion of the steel structure.

Disposal

Nothing lasts for ever. How is the building going to be disposed of at the end of its life? Can the material be re-used or converted into some other usable form? What are the costs associated with this?

Fire protection

There is an unfortunate possibility that any structure may catch fire. Some materials have better fire resistance properties than others.

Size and nature of the site

The location of the site may influence the choice of materials. Traffic congestion problems, local by-laws and physical obstructions may limit the size of deliveries to the site and the times of day that deliveries can take place.

We will now discuss each of the main structural materials individually. As you will see, each material has its advantages and disadvantages.

Concrete

Concrete is manufactured by mixing four ingredients – cement, fine aggregate (sand), coarse aggregate (gravel or crushed rock) and water – in pre-determined proportions in a controlled manner to form a grey fluid resembling porridge. This wet concrete is transported to the place where it is needed and poured into 'moulds' of the required shape and size. These moulds, known by the terms *formwork* or *shuttering*, are usually made of timber or steel. Chemical reactions take place within the concrete, which lead to its setting, hardening and gaining in strength over a period of weeks.

The production of concrete needs to be carefully controlled. Firstly, its naturally occurring constituent materials are variable in quality. Secondly, wet concrete is susceptible to high or low temperatures and needs to be placed as quickly as possible before it 'goes off'. Thirdly, careless treatment of wet concrete – for example, allowing it to drop from a great height or to bounce off formwork – can lead to segregation of its constituents which can affect the integrity of the finished concrete.

Concrete is strong in compression (typically 30–40 N/mm^2) but weak in tension (3–8 N/mm^2). As we saw in Chapter 3, any structural element in bending – for example, a beam or a slab – experiences tension, therefore, if made of concrete, it needs to be reinforced with steel bars. Concrete with steel bars in it is known as *reinforced concrete*. In practice, all concrete seen in structures is reinforced concrete.

Reinforced concrete has a number of advantages:

- It has high strength when reinforced.
- It is mouldable into any desired shape.
- Because it is mouldable, it can be formed into structurally continuous elements.
- It is durable: it does not corrode or rot.
- It has good fire resistance properties.
- It also has good thermal and noise insulation properties.

- It is relatively cheap to produce – although its placement on site is quite labour-intensive, which increases the cost.
- It can be used compositely (that is, two materials acting together) with structural steel.
- It is widely used in foundations, columns, beams, slabs, bridges, roads, railway sleepers.
- It is suitable for short-span low- and high-rise building frames.
- Prestressed concrete – concrete through which highly tensioned rods or cables have been placed – is stronger than reinforced concrete and therefore longer and more slender members can be produced. Prestressed concrete is therefore suitable for long spans and rigid frames.
- Concrete elements (beams, columns, etc.) can be made in factories and then, when hardened, transported to a construction site and erected into position. Such elements are termed *precast* concrete elements. The more usual concrete construction, where wet concrete is poured into formwork on site, is called *in-situ* construction.

However, the following disadvantages of reinforced concrete also need to be considered:

- It is heavy, both physically and aesthetically.
- As indicated above, construction using reinforced concrete needs to be carefully controlled and is labour intensive. It is 'messy', requiring formwork, reinforcement and the placing and compaction of concrete.
- Once poured, it takes several weeks for the concrete to achieve the required strength. This delays consequent construction activities (unless the concrete is precast).
- Although it doesn't rot or corrode, concrete can suffer certain ills, including spalling, cracking (leading to possible corrosion of reinforcement) and carbonation (a chemical reaction with the atmosphere that causes deterioration).

Masonry

Traditionally the term masonry refers to the material crafted by a mason – namely, stone. In modern times, the term more usually applies to brickwork or blockwork.

Bricks and blocks come in small, cuboidal units which can be lifted by hand. They are laid in rows by a bricklayer to form walls or columns. Mortar is used to 'glue' the individual units together and to fill the gaps or any irregularities between units.

The advantages of masonry are as follows:

- It has high compressive strength, making it ideal for walls, columns and arches, all of which are in pure compression.
- It is durable – no finish is required.
- It is made from raw materials readily available in the UK at low cost.
- No complicated plant is required.
- It has an attractive appearance.
- There is design flexibility – bricks or blocks can be combined to form complex shapes.
- Masonry has good fire resistance properties and good thermal/acoustic properties.

The disadvantages of masonry are:

- It has very low tensile strength, which means it cannot be used for elements which bend, for example beams or slabs.
- Compared with timber (the other material used for low-rise domestic construction), masonry is heavy, so larger foundations are required and transport costs are higher.
- Frost and chemical attack can cause spalling in brickwork.
- Efflorescence – chalky and unsightly (but harmless) deposits – can occur on brickwork following a cycle of wetting and drying.

Figure 21.1 shows a traditional stone arch bridge. Arch structures are in compression throughout and stone, being strong in compression, is an excellent material for such structures.

Fig. 21.1 Stone arch bridge.

Timber

Timber is the only structural material which is used in its naturally occurring form. The length and cross-section of a timber beam are limited by the height and girth of the tree from which it is obtained.

Longer timber beams, and larger cross-sections, can be obtained by slicing the timber into thin strips and gluing these strips together both along their lengths and their ends, but it is an expensive process rarely used in the United Kingdom. This is known as glued laminated (or 'glulam') timber.

Timber comes in two types:

- hardwoods, obtained from deciduous (leaf-shedding) trees;
- softwoods, obtained from coniferous (evergreen) trees.

Softwoods are generally used for structural purposes. Strength of a given species of timber is determined either visually (from inspection by a suitably skilled person) or mechanically (by laboratory testing).

Timber is one of the oldest building materials and has the following structural advantages:

- It is light, with a high strength/weight ratio.
- It is easy to cut and shape.
- Despite what you might expect, it performs well in fire.
- It has good chemical durability.
- It has a pleasing appearance.
- It is relatively cheap.
- Although it has low stiffness, it is relatively stiff in relation to its own (light) weight.
- It is suitable for lightly or moderately loaded low-rise building frames and for shed and rigid frames.

But timber has the following disadvantages:

- Its low strength means that spans are limited, as is the height of timber buildings.
- It is difficult to form joints in certain circumstances.
- As mentioned above, the size of a piece of timber is limited by the size of the tree from which it comes.
- Timber is susceptible to rot and decay unless properly maintained.
- Its properties vary according to species of tree.

Steelwork

Structural steelwork is manufactured in standard sections. It has the following advantages:

- Its strength is high in both tension and compression (but compression in steelwork can be a problem – see below).
- Steel has a high strength/weight ratio.

- Because steel sections are produced in a factory under carefully controlled conditions, high quality control can be achieved.
- Steel's appearance can be elegant, with slender elements, smooth surfaces, straight and sharp edges.
- Pre-fabrication is possible.
- Steel has high stiffness.
- Steel is economic in material: a small amount carries a relatively large load.
- Steel is suitable for low/high-rise buildings and roof structures of all spans.

Steelwork does, however, have the following disadvantages:

- It is difficult to form curves.
- It is heavy: cranes are required to lift steelwork.
- It is a high-cost material.
- It has a durability problem: it corrodes if not protected and maintained.
- It has poor fire resistance; therefore steelwork needs to be protected by other materials.
- Because of the slender sections used in steelwork, it is prone to buckling in compression. This is an important criterion in the design of steelwork.

The complex steel and glass footbridge structure shown in Figure 21.2 was constructed in the late 1990s to connect two shopping centres.

Fig. 21.2 Steel footbridge, Manchester.

Aluminium

Aluminium is rarely used as a structural material except in very small structures (e.g. greenhouses). Its main properties are as follows:

- Its strength is about the same as mild steel.
- It is stiffer than concrete or timber.
- It is less stiff than steel, but also lighter.
- It has high strength/weight ratios.
- But: aluminium is expensive.

So how do I decide what materials to use in a given building?

The following discussion relates to construction in the United Kingdom, though some of it may apply elsewhere.

Framed or unframed structure?

The first decision to be made is whether the structure will be framed or unframed. In a framed structure, a framework or 'skeleton' of beams and columns is used to carry the structural loads down the building to the foundations. The framework is usually of steel or reinforced concrete, but in very small (usually single-storey) structures may be of timber or aluminium. The finished building will usually also have external and internal walls, but these are non-structural and support no loads other than their own weight.

In a non-framed structure, the walls are load-bearing. These load-bearing walls are usually masonry, but may be reinforced concrete.

Example 21.1

Consider the following scenario.

Depending on your specialism, you run either an architectural practice or a firm of consulting engineers. One of your clients, a property development company, proposes to construct an office development on a specific site. Dimensions of the planned building have yet to be finalised, but it is known that the building will be two-storey, of approximate plan dimensions 60 m x 20 m. When complete, the building will be rented out to either one company or, with appropriate subdivisions, to a number of small tenant companies.

At the first meeting of the project team, your client asks your advice on whether a framed structure would be appropriate. Write your reply, giving full reasons for your choice.

Having thought about this, your answer would probably be that a framed structure is the appropriate option, for the following reasons:

- It is clear that the use of the building is not rigidly defined. It is an office building, but may be occupied by a number of companies, and the tenant companies may grow (thus requiring more space) or shrink (requiring less space). Companies may come and go over time. Accordingly, the available space should be as flexible as possible to accommodate the changing needs of the tenants. It is best not to have such flexibility inhibited by the presence of internal load-bearing walls.
- The absence of load-bearing walls means there will be more floor space. Although this increase in floor space will be relatively small, it will be good news for your property developer client, who will be anxious to squeeze as many lettable square feet as possible out of the building.
- If there are no load-bearing walls – which would be made of concrete or masonry and so would be relatively heavy – the building as a whole will be lighter. This relative lightness would mean that the loads on the foundations would be less, which in turn means that the foundations could be less substantial and thus cheaper. Your client would be delighted at any saving in money that you could offer him.
- Framed structures of steel or concrete can be erected much faster than load-bearing masonry structures. This will again please your client, who will want to see the structure completed (and thus providing rental income) as soon as possible – preferably yesterday.

However, as with most projects in 'the real world', things do not run smoothly and there is a twist in the tale:

At the second meeting of the project team, your client shares his belief that a forthcoming recession will cause a drastic decrease in the demand for office accommodation. He does, however, foresee a growing demand for quality hotel accommodation and has therefore replaced the office project with a hotel project on the same site, which, when complete, will be sold to the Dream Easy Inn hotel chain for use as a bedroom block. Due to planning constraints, the height and overall dimensions of the building will remain as before.

Your client asks whether this change of use would change your earlier advice on the building's structure. What is your reply? Give reasons.

Now the scheme has changed totally. Although the final building will be the same shape and size as before, its use is now completely different. The needs of a hotel chain (and the guests who pay to stay there) are vastly different to the demands of a company renting office space (and those of the office workers it employs). So the architect and engineer need to think again.

In this case, you may well decide that a framed structure is not appropriate, for the following reasons:

- Guests in a hotel room want a good night's sleep. It is therefore important that the hotel room be at the right temperature and quiet – no guest wants to be disturbed by noise from the room next door or from outside. High levels of thermal and sound insulation are therefore important. It makes sense to use load-bearing blockwork which, correctly specified, would provide an appropriate level of thermal and sound insulation as well as forming part of the building's structure.
- Unlike the office scenario, no flexibility is required of a hotel bedroom block. It is unlikely that the hotel owner would need to change the size of individual hotel rooms or the location of their walls in the future.
- Once again, you should consider your client's needs. As he will be selling the building on to a hotel chain on completion, his main concern is that the finished building will be an attractive purchase for such an operator. Your client is not concerned about the building's future income potential.
- It should be noted that this building is low-rise (only two storeys). The decision might be different with a high-rise building, where the efficiency of a structural framework would override other considerations.

We can extrapolate the lessons we've learned from this specific example to more general cases, as follows.

Features of framed structures:

- flexibility: can accommodate change of use;
- small saving in floor space;
- lighter, giving smaller (and hence cheaper) foundations;
- faster speed of erection.

Features of non-framed structures:

- inherent thermal and sound insulation properties in masonry, so useful for hotels or apartment buildings where insulation is important;
- no flexibility in the use of the building – but this may not be required anyway.

The following is a list of the materials used for particular structural elements.

Walls
- Masonry (unframed structures).
- Masonry, timber stud, aluminium frame (framed structures).

Floors
- Timber joists supporting floorboards (domestic: low loads, small spans).
- In situ reinforced concrete (general industrial/commercial).
- Precast concrete (suitable for regular, repetitive floor layouts).
- Composite: in situ concrete on corrugated steel (popular for office buildings).

Beams
- Timber (short spans only).
- In situ reinforced concrete (general industrial/commercial).
- Precast concrete (not common unless prestressed).
- Prestressed concrete (suitable where long spans are required).
- Steel.

Columns
- Timber (domestic and other small-scale construction only).
- Reinforced concrete.
- Steel.

Pitched roofs
- Timber truss or rafter/purlin construction (domestic only).
- Steel truss or portal frame (longer-spanned commercial/industrial buildings).

Foundations
- Concrete (usually reinforced for other than domestic construction).

More on materials

Material selection for structural design

In earlier chapters of this book we have looked at such matters as shear force, bending moment and stress. We have learned how to evaluate these things and, in the case of shear force and bending moment, draw diagrams of their distribution. Some readers may have wondered how we apply this information. For example, we might calculate that the maximum bending moment experienced in a particular beam is 45 kN.m, or that the compressive stress in a certain column is 25 N/mm^2, but how do we make use of this information?

The process of converting a piece of information such as maximum bending moment = 45 kN.m to a reinforced concrete or steel beam of a shape and size that will resist this bending moment is known as *structural design*. The full structural design process is beyond the scope of this book – there are many excellent textbooks available on the subject – but this chapter serves as an introduction to structural design.

The first decision that the structural designer needs to make is what material – or combination of materials – should be used in a given situation. In Chapter 21 we discussed the four main materials used in structural design (steel, reinforced concrete, masonry and timber), the advantages and disadvantages of each, and which material(s) is likely to be used for any particular type of structural member. This should guide you in your material selection. We will now discuss the alternative forms of building construction that are available to the designer.

Alternative forms of construction

The most common types of structural schemes for buildings are outlined in the following sections.

Steel frames

These are structural frameworks comprising steel beams and columns supporting floor slabs. The floor slabs are usually of concrete or a steel/concrete composite such as profiled steel decking onto which concrete is poured. The beams that span between columns are called *primary* beams and they in turn may support *secondary* beams (we saw an example of this in Chapter 5).

Lateral stability of the structure is important and we saw in Chapter 11 that this may be assured by using diagonal cross-bracing or by designing the beam/column joints to be sufficiently rigid. Such measures may also be necessary to prevent torsion (twisting) of the building.

Steel beams and columns are available from manufacturers in standard section sizes and tables of the structural properties of these standard sizes are available for designers. While deeper sections may be stronger and lighter (and hence, in material terms, cheaper) than shallower ones, headroom considerations may lead to the overall building height being greater if deeper sections are used. This will lead to increased costs because the increased height of the building means that a greater number of columns, cladding, lifts, etc. is required.

Services (that is, electrical and telephone cables, gas and water pipes and ventilation ducts) need to be accommodated. We will see in Chapter 23 that some types of steel beam can cater for such services more readily than others.

Steel beams and columns need to be connected to each other, usually by bolting or welding. Connection details need to be kept simple in order to keep the costs (of fabrication, installation and material) to a minimum. And, as we've already seen, steel is vulnerable to fire and corrosion and needs to be protected accordingly. Because steel sections are slender, they are also vulnerable to buckling, a consideration that needs to be addressed at the design stage.

Figure 22.1 shows a typical steel-framed office building under construction.

Reinforced concrete frames

These are frames of concrete beams and columns. As we saw in Chapter 21, structural concrete is always reinforced internally with steel bars in order to provide the required tensile strength. Reinforced concrete frames usually comprise in situ concrete: this means that the concrete beams or columns are formed by pouring wet concrete into a mould (formwork) located at the beam or column's final position. The formwork needs to be supported by a temporary propping structure which, along with the formwork itself, needs to be left in position for several days until the concrete has gained sufficient strength. This requirement can impede and delay other site activities.

Fig. 22.1 Steel-framed office building under construction.

Reinforced concrete beams support reinforced concrete slabs, which can be of various types, for example, one-way spanning, two-way spanning, rib or waffle, which were illustrated in Chapter 3. Unlike steel structures, fire protection is not normally a problem with concrete and, provided cracking is kept within acceptable limits, neither is corrosion of the steel reinforcement.

Construction of reinforced concrete buildings is quite labour-intensive: operatives are required to make and install the formwork and its supports, place the reinforcement and place the concrete.

Precast concrete frames

Precast concrete frames comprise individual beams and columns of reinforced concrete that have been made in a factory then delivered to site in their completed form. (This contrasts with the in situ concrete frames discussed above, where the concrete is formed at its final position on site.) Greater quality control can normally be achieved with precast members, as the environment in a factory is more controllable than that on site. Also, as the precast members will have achieved their full strength when they arrive on site, there will be no waiting time to delay other site activities. However, precast construction is best suited to structures which are totally regular and repetitive in nature.

Timber frames

Because of timber's limited strength, timber-framed structures tend to be small-scale domestic buildings. Timber roof frames are generally used for domestic pitched roofs. Connection between members, and protection of timber against rot or insect attack, is also a consideration.

Load-bearing masonry

Structural masonry – stone, brickwork and blockwork – is the preferred form of construction for the walls of houses and other non-framed structures in the United Kingdom and elsewhere. Masonry's high compressive strength makes it ideal for such structures and also for other structures that are in pure compression: for example, arches. Masonry comes in small units (e.g. individual bricks) which are easy to manage. However, skilled labour is required.

Masonry is generally less tolerant to differential settlements and accidental damage than steel or in situ concrete-framed buildings.

Hybrid schemes, e.g. steel frame with precast concrete floors

These are combinations of the above.

The choice between different construction types

The choice between construction types will depend on the following factors.

The need for flexibility

As we saw in Chapter 21, the future use of the building – and whether this is likely to change over time – may influence the type of construction.

The spans required

Sometimes there are requirements for long uninterrupted spans in, for example, theatres and other auditoria, multi-storey car parks, exhibition halls or, in the case of bridges, shipping lanes. In such cases the span will usually dictate the form of construction. As a general rule, the longer the crossing, the more expensive it will be to achieve.

The ground conditions

These will dictate whether relatively cheap conventional foundations can be used or whether more expensive piling or other foundation types are required.

The access to the site

If the site is isolated, access roads may need to be built. On the other hand, if the site is in the middle of a large city, delivery of materials may be restricted. There may be other constraints, for example, all materials may have to be lifted in by crane because of physical site constraints. The wise designer considers all this at the design stage.

The experience of the designers

The designers may have great experience of a particular type of design which can be used on the present project. This makes the whole process less painful – and less costly – because of the benefits of the lessons learned on previous projects.

The experience of the contractors

Again, the contractors may have experience of certain construction types and techniques which, if used on the present project, will keep the cost down.

The availability of materials

It is no use specifying materials which are either unavailable or have to be imported at great cost, whatever their attributes might be.

Risks and difficulties in the construction process tend to lead to increased costs, so the chosen solution should seek to minimise these. Once you have decided on the material (or combination of materials) to use, the design process can begin.

Design from first principles and design standards

You can learn how to design (for example) a reinforced concrete beam from first principles. This largely mathematical process is taught at some universities and is dealt with in some textbooks. But bear in mind that you aren't the first person to attempt to design a reinforced concrete beam and you won't be the last. The problems you encounter in doing so have all been encountered before and attempts have been made to deal with the process in documents called Standards or Codes of Practice. In the remainder of this discussion I will assume that you are either in the United Kingdom or some other places where British Standards are used, as my discussion will be based on British Standards. However, the general principles of what I say will apply to other Standards (e.g. the American ASTM Standards) also. In addition, you should be familiar with the local building codes of the country or place where you are, as local building codes overrule the requirements of Standards.

British Standards

Are British Standards a set of rules and regulations, legal documents or design guides? Certainly, all design must conform to the requirements of the relevant British Standard, but to what extent do British Standards fulfil the role of design guides? Certainly, a person who is given half an hour to design a reinforced beam is unlikely to find the relevant British Standard much use unless he or she has been trained in putting it into practice. However, a person familiar with the Standard will be able to use it to design a beam very quickly.

The early British Standards may have been user-friendly design guides. However, British Standards have become more and more voluminous over the years and, in my opinion, now read more like legal documents.

An individual new to structural design needs guidance in the use of British Standards. If you study a course or module entitled 'structural design' (or similar) at university or college, your lecturer or tutor will (or should) act as a friendly guide through the relevant parts of the code. Some textbooks are also good at performing this role.

The good news is that the design of, say, a timber beam or a structural steelwork column is a fixed procedure which can be easily followed or learned – effectively, once you've designed one masonry wall you've designed them all!

The relevant British Standards and where to find further information

The relevant British Standards are:

- BS 8110: Part 1: 1997: Structural use of concrete.
- BS 5628: Part 1: 1992: Structural use of unreinforced masonry.
- BS 5268: Part 2: 1996: Structural use of timber.
- BS 5950: Part 1: 2000: Structural use of steelwork in building.

I'm sure it's totally unintentional on the part of the drafters of the British Standards, but you'll realise how easy it is to confuse the masonry and timber standards with each other (BS 5628 and BS 5268 respectively).

Eurocodes will replace these British Standards within the next few years. These documents will be used for design throughout the European Union and it is hoped that this international consistency will make it easier for engineers from different countries to work together. Each Eurocode should be read in conjunction with the relevant National Application Document (NAD) which will give parameters which should be used for specific countries. The relevant Eurocodes are listed below:

- EN 1990 Basis of structural design
- EN 1991 Actions on structures
- EN 1992 Design of concrete structures
- EN 1993 Design of steel structures
- EN 1994 Design of composite steel and concrete structures
- EN 1995 Design of timber structures

- EN 1996 Design of masonry structures
- EN 1997 Geotechnical design
- EN 1998 Design of structures for earthquake resistance
- EN 1999 Design of aluminium structures

Apart from the first, each of these Eurocodes is divided into a number of parts. (Note: The numbering of these Eurocodes is unfortunate and should not be confused with years. For example, EN 1996 has nothing to do with the year 1996.)

How far can I span?

Introduction

Any person involved in the conceptual design of a building will soon have to consider how far can be spanned in practice. There are no easy answers to this question (although some rules of thumb are given below), but this chapter explores the various factors involved.

Long span structures

What is a span?

You will be most familiar with the word *span* in the context of bridges, but it applies to beams and slabs within building structures as well. The span of a bridge (or beam, or whatever) is the horizontal distance between supports.

How far can we span?

On the face of it, a span can be as long as is necessary. A concrete lintel struggles to span a 1 metre wide door opening, while modern suspension bridges can – and do – span several kilometres. In practice, the greater the span, the stronger the spanning element has to be. This generally means it has to be deeper and this greater depth may be difficult to accommodate physically. And inevitably, as span increases, the cost will also increase.

However, spans should not be too small either, as an excessive number of columns or supporting walls can interfere too much with activities going on within the building – nobody wants to see a 'forest' of columns. In general, spans are made as long as is reasonably practical.

Some points worth noting are as follows:

250 Basic Structures for Engineers and Architects

Fig. 23.1 Building with cantilevered canopy, Berlin.

- Some materials are stronger than others, therefore can span greater distances.
- In general, the longer the span, the deeper the supporting element has to be.
- Some building uses dictate that large uninterrupted spans are required. Examples include sports halls, swimming pools, theatres, concert halls, etc.

At first glance, the canopy shown in Fig. 23.1 looks impossibly wafer-thin in respect of the distance it is cantilevering. However, Fig. 23.2 shows a break in the canopy that reveals it is in fact considerably deeper (and hence stronger) at its supporting columns. Careful design conceals this fact elsewhere from street-level observers.

Let's consider the various structural materials one by one.

Steel

Standard steel beams

Steel beams (universal beams) are manufactured in standard sizes. Corus (formerly British Steel) produces tables of these standard sizes and their various dimensions and properties. In many cases a steel building will comprise standard beams (and columns), selected by the designer to accommodate the calculated bending moments, shear forces and deflections. A typical universal beam section is shown in Fig. 23.3 (a).

Fig. 23.2 Building with cantilevered canopy, Berlin (detail).

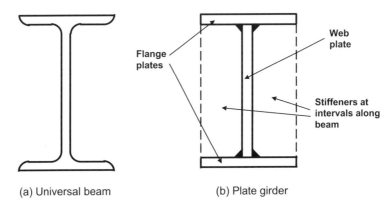

(a) Universal beam   (b) Plate girder

Fig. 23.3 Universal beams and plate girders.

Plate girders

The largest standard section Corus produces is 914 mm deep. In situations where even this largest size is not adequate, it is possible to fabricate larger sections by welding together plates in the right configuration. In other words, a section can be 'tailor made' when no 'off the peg' section is adequate. A typical plate girder section is shown in Fig. 23.3 (b).

Castellated and cellular beams

These steel beams have large holes in their webs (that is, the vertical parts) at regular intervals along the beam. These holes are hexagonal in the case

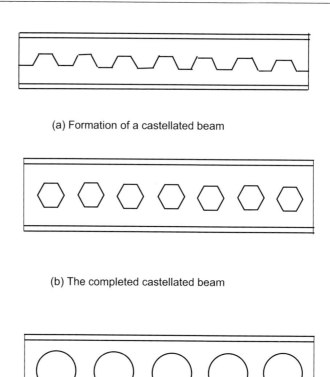

(a) Formation of a castellated beam

(b) The completed castellated beam

(c) A cellular beam

Fig. 23.4 Castellated and cellular beams.

of castellated beams and circular in the case of cellular beams. Cellular beams in particular are very popular in multi-storey steel-framed construction. Castellated beams are formed by cutting a standard steel beam longitudinally along a zigzag, as shown in Fig. 23.4 (a), then reconnecting the two half beams as shown in Fig. 23.4 (b), thus forming a deeper (and therefore stronger) section of the same weight as before. Moreover, the holes in castellated and cellular beams can be used to accommodate services such as cables or water pipes.

Lattice girders

But what happens when a plate girder section would have to be so big as to be impractical? Well, instead of using a solid steel beam (as shown in elevation in Fig. 23.5 (a)), a lattice girder beam could be used. As you will recall, the top of a sagging beam is in compression and the bottom is in tension. A lattice girder comprises a top boom (in compression) and a bottom boom (taking the tension), with the two booms linked by diagonal members.

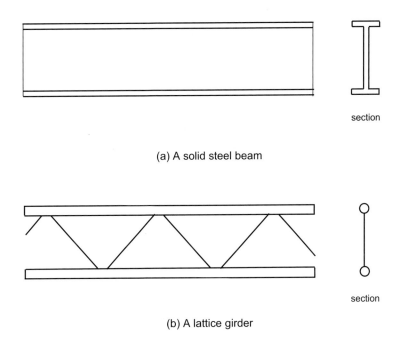

Fig. 23.5 Solid steel beams and lattice girders.

Using a lattice girder, a deep beam can be achieved without the requirement for it to be solid. This saves on material (and thus weight) and means that the gaps within the lattice can be used for other things (e.g. services can pass through them). A typical lattice girder is shown in Fig. 23.5 (b).

Lattice girders (referred to there as 'bar joists') are a common type of floor construction in commercial buildings in North America. These bar joist are typically 300–400 mm deep and are spaced at (typically) 600 mm.

Lattice box girders

What happens when the required span and loading increase still further? We could continue deepening (and thus strengthening) the lattice girder. Another option is to introduce a second lattice girder running alongside the first one and linked to it by two horizontal lattice girders, one at top boom level, the other at bottom boom level. A box is thus formed and therefore this type of beam is called a box lattice truss, shown in Fig. 23.6 (a). A variation on this theme is the triangular lattice truss, shown in Fig. 23.6 (b). As steel is more likely to buckle when in compression, the cross-section of steel available in the compression zone is maximised by having two booms in this zone and one in the bottom (tensile) zone as shown. Figure 23.7 shows triangular steel lattice trusses, double curvature in profile, spanning an airport concourse and supporting its roof.

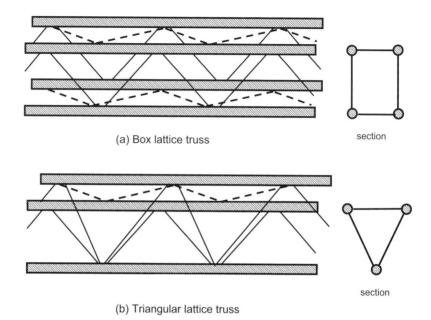

(a) Box lattice truss

(b) Triangular lattice truss

Fig. 23.6 Lattice trusses.

Fig. 23.7 Triangular lattice trusses supporting terminal roof, Liverpool John Lennon Airport.

Suspension structures

Lattice box girders can span considerable distances – typically up to 100 metres when supporting football stands – but there may be cases when we need to span greater distances. In such cases we need to use suspension or cable-stayed structures. The principle behind these is that if support cannot be provided from below, it can be provided from above by means of cables which run over supporting masts to an anchorage point in the ground.

Concrete

Reinforced concrete

As discussed in Chapter 21, structural concrete is always reinforced – i.e. it has steel bars embedded in it – for strength purposes. Unfortunately, the span/depth ratios required of reinforced concrete are not very desirable to designers: if a reinforced concrete beam is required to span a long distance, its depth will be inconveniently great.

Prestressed concrete

Prestressed concrete beams (i.e. those containing embedded steel bars or cables subjected to large tensile forces) can be much more slender than reinforced concrete beams and therefore are a popular choice when long spans are required. Prestressed concrete beams are often visible in multi-storey car parks.

Timber

Timber beams and joists

The material aspects of timber were discussed in Chapter 21, where we saw that timber beams cannot span great distances because of their limited strength. The cross-sectional size is also limited by the size of the tree from which the timber was obtained.

Glued-laminated ('glulam') beams

Longer spans are possible with timber if glued laminated beams are used. As mentioned in Chapter 21, such beams are formed by building up layers from thin slices of timber glued together.

Glulam beams are not common in the United Kingdom due to their high cost, but are sometimes seen supporting the roofs of swimming pools – timber being less susceptible to the corrosive action of chlorine gas than other materials.

Masonry

As discussed in Chapter 21, masonry is weak in tension and is therefore not really suitable as a spanning material. This is why the columns in ancient Greek and Egyptian temples are so close together: the stone beams that span between them can span only a short distance.

Masonry arch structures

Masonry, being strong in compression, is suitable for use in arch structures, which are in compression throughout. Masonry arches can span reasonable distances and a series of masonry arches can be used to form a viaduct, as can be seen in Roman stone aqueduct structures (for example, at Nimes in the south of France) and in Victorian brick railway viaducts at many locations in the United Kingdom and elsewhere.

Spans and depths: some rules of thumb

A question I am commonly asked by students of architecture is: 'How far can I span and how deep would the beam have to be?' If only it were that simple.

As mentioned earlier, in broad terms, the greater the span, the greater the depth. It follows that rule of thumb span-to-depth ratios can be generated and these are given in Table 23.1. These should be used with caution and the following points should be noted:

- The possible spans, and associated depths, depend on the loading to which the beam is subjected. The figures in Table 23.1 assume 'normal' commercial building loads. They do not apply to more heavily loaded situations (e.g. plant rooms) or to unconventional loading scenarios.
- This information is given without prejudice and is for guidance purposes only. It is suitable for initial sizing of structural elements for architectural scheme or costing purposes.
- For actual building projects the size of structural elements must be verified through detailed design by a qualified structural engineer.

Table 23.1 Span ranges and span/depth ratios

Type of element	Span range (metres)	Typical span/depth ratio
CONCRETE		
Beam: simply supported	Up to 8 m	15–20
Beam: continuous	Up to 12 m	20–27
Beam: cantilever	Up to 5 m	1–7
Slab: one-way: simply supported	Up to 6 m	20–30
Slab: one-way: continuous	Up to 6 m	20–30
Slab: one-way: cantilever	Up to 3 m	5–11
Slab: two way: simply supported	Up to 6 m	30–35
Slab: two way: continuous	Up to 6 m	30–35
Profiled steel decking/concrete composite	Up to 6 m	35–40
Ribbed slab	Up to 11 m	35–40
Waffle slab	Up to 15 m	18–25
Column	Storey height	10–17
Strip foundation	0.8–2.0 m wide	
Pad foundation	1.5–3.0 m square	
TIMBER		
Joist flooring	Up to 6 m	10–20
Glulam beam	Up to 30 m	15–20
Plyweb beam	Up to 20 m	10–15
STEEL		
Primary beams (supported by columns)	Up to 12 m	15–20
Secondary beams (supported by other beams)	Up to 7 m	15–20
Portal frame	Up to 60 m	35–40

Calculating those loads

Introduction

In the earlier chapters of this book you were shown how to calculate such things as shear force, bending moment and stresses. In worked examples you were presented with loads to work with. Unfortunately, real-life structural problems are not as neatly packaged as examples you might encounter in textbooks or in university lecture theatres. You have to calculate the loads yourself. This chapter tells you how.

As you learned in Chapter 5, there are two types of loading:

(1) dead (or permanent) loads;
(2) live (or imposed) loads.

We will discuss the calculation of each of these in turn, then look at some examples.

Dead load

Unit weights of common building materials are given in Appendix 1. (British Standard BS 648 gives the unit weights of a much wider range of materials.) These loads are expressed in kN/m³ and represent the weight of a cubic metre of the material. For example, the unit weight of reinforced concrete is 24 kN/m³, which means that a cubic metre of concrete weighs 24 kN. This is almost two-and-a-half times the weight of water. So if, in the early stages of your career, you have to carry buckets full of wet concrete short distances on construction sites (as I did), you'll find they're considerably heavier than the buckets of water you use when cleaning your car!

If you want to calculate the weight of a reinforced concrete beam which is 200 mm (or 0.2 metres) wide, 400 mm (or 0.4 metres) deep and 6 metres long, you first need to calculate the volume of the beam, then multiply it by the unit weight to get the total weight.

Volume of beam = length × breadth × height
= 6 m × 0.2 m × 0.4 m
= 0.48 m^3

Total weight of beam = volume × unit weight
= 0.48 m^3 × 24 kN/m^3
= 11.52 kN.

Live load

As you will recall from Chapter 5, these are loads due to people and furniture. By their very nature, these are variable. To simplify matters, live loads are assigned certain values depending on the use of the building or area concerned. These loads are expressed in kN/m^2 and typically fall in the range 1.5–5.0 kN/m^2. Values for some common cases are given in Appendix 1.

For example, the live load relevant to classrooms is 3.0 kN/m^2. So, for a classroom which has a floor area 10 metres × 10 metres:

Total live load = 10 m × 10 m × 3.0 kN/m^2
= 300 kN.

Example 24.1: Loading on a reinforced concrete beam

A reinforced concrete beam spans 6 metres between supporting columns. The beam is 250 mm wide and 450 mm deep and supports a 5 metre wide portion of slab 175 mm deep. There is a 40 mm deep concrete topping layer on top of the slab. The floor supports offices. There is also a non-loadbearing masonry (blockwork) wall directly above the beam and running along the line of the beam. This blockwork wall is 2.5 metres high, 200 mm thick and is finished on both sides with plaster of weight 0.5 kN/m^2. See Fig. 24.1.

Calculate the total load on the concrete beam per metre length. (Note: don't forget to include the weight of the beam itself.)

Solution

Unit weight of concrete = 24 kN/m^3. (Appendix 1)

Unit weight of blockwork = 22 kN/m^3 (Appendix 1)

Live load due to offices = 2.5 kN/m^2 (Appendix 1)

Note that you are asked to calculate the total load per metre length of concrete beam. There will be a number of contributions to this total load. They come from the blockwork wall, the plaster on it, the topping layer on top of the slab, the slab itself, the beam's own weight and the live load (due to people and furniture) on the slab. One of the basic

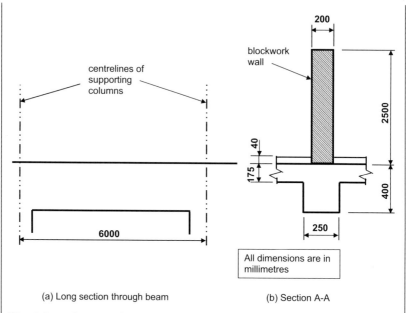

(a) Long section through beam   (b) Section A-A

Fig. 24.1 Concrete beam featured in example 24.1.

mistakes students sometimes make with this sort of calculation is in forgetting one or more of these contributions.

Considering these contributions one by one:

Blockwork wall:

This is simply a matter of multiplying the volume of the wall (length × breadth × height) by its unit weight:

Load due to blockwork wall = 2.5 m × 0.2 m × 1.0 m × 22 kN/m^3
= 11 kN

Plaster on blockwork wall:

The unit weight of plaster has been expressed in units of kN/m^2 – in other words, load per unit area. This means that we must calculate the total plastered area per metre length of wall and multiply this area by the unit weight of plaster given above. Remember, the wall is plastered on both sides, so the number 2 in the calculation below represents two sides:

Load due to plaster = 2 × 2.5 m × 1m × 0.5 kN/m^2 = 2.5 kN

Concrete slab:

As with the blockwork wall, we multiply the volume of the slab (per metre length of beam) by the unit weight of reinforced concrete. Remember: the beam is supporting a 5 metre wide portion of slab.

Load due to reinforced concrete slab = 5 m × 0.175 m × 1 m × 24 kN/m^3
= 21 kN

Topping layer:

It is reasonable to assume that the unit weight of the topping layer is the same as that for the structural reinforced concrete. As with the slab proper, we multiply the volume of the topping by the unit weight. To make calculations easier, we shall pretend that the topping continues underneath the blockwork wall – even though it doesn't. This means our calculation will be slightly conservative (i.e. an over-estimate).

Load due to topping layer = 5 m × 0.04 m × 1 m × 24 kN/m^3 = 4.8 kN

Concrete beam:

We have already considered the top part of the beam (i.e. the top 175 millimetres of the beam's depth) when we were calculating the load due to the slab. We now need to calculate the loading due to the bottom 225 mm (i.e. 400–175) of the beam.

Load due to reinforced concrete beam = 0.175 m × 0.250 m × 1m
× 24 kN/m^3
= 1.05 kN

Live load:

This was expressed above as a load per unit area of floor slab (2.5 kN/m^2). Again, for simplicity we'll ignore the presence of the blockwork wall when calculating this load. The live load will be the surface area of the slab multiplied by the load per unit area, as follows:

Live load = 5 m × 1 m × 2.5 kN/m^2 = 12.5 kN

So:

Total dead load = (11 + 2.5 + 21 + 4.8 + 1.05) = 40.4 kN

Total live load = 12.5 kN

This gives a total load of 52.9 kN per metre length of the beam.

Example 24.2: Loading at the base of a column

A four-storey reinforced concrete-framed building has a plan area of 18 metres × 25 metres. Supporting columns are arranged on a grid of 6 metres × 5 metres, as shown in Fig. 24.2. At each level, floor slabs span 5 metres onto supporting beams, which in turn span 6 metres between columns. The 6 metre span beam considered in Example 24.1 is a typical supporting beam.

If the ground floor slab is ground-bearing (in other words, it is supported directly on the ground underneath) and the live load on the flat roof is the same as on the floors, calculate the total load at the base of a typical internal supporting column if the columns are 400 mm × 400 mm in plan.

Fig. 24.2 General arrangement of 4-storey office building.

Solution

At each level, a typical column will support an area of beam and slab as shown by the hatched zone in Fig. 24.2. In Example 24.1 we have already calculated the total load on a typical 1 metre length of beam, so if we multiply this figure by 6 metres, we have the total load supported by a typical column at each level. We then need to multiply the result by 4 to represent the four floors (excluding the ground-bearing ground floor slab but including the roof slab).

Total load on typical column from beams = 52.9 kN/m × 6 m
× 4 storeys
= 1270 kN

We now need to add on the weight of the column. If the total height of the building (and thus of the column) is 14 metres, the weight of the column is:

Column self-weight = 14 m × 0.4 m × 0.4 m × 24 kN/m² = 53.8 kN

(Again, this is obtained by working out the volume of the column and multiplying it by the unit weight of concrete.)

So:

Total load at base of a typical internal column is: 1270 + 53.8
= 1323.8 kN

Example 24.3: Sizing of a pad foundation

Typically, a column in a building will be supported by a pad foundation. The function of a pad foundation – indeed, any foundation type – is to transmit the loads from the building's superstructure (that is, the above ground part) safely into the ground beneath.

In order to determine a foundation size, two things need to be known:

- The total load on the foundation.
- The permissible ground-bearing pressure.

The permissible ground-bearing pressure – in other words, the maximum pressure that the ground can sustain without deforming – can only be determined from a ground investigation relating to the site of the proposed building.

If a ground investigation has been done for the site of the building discussed in Example 24.2 and the permissible ground-bearing pressure has been found to be 200 kN/m², calculate the pad foundation size required.

Solution

$$\text{Minimum pad size required} = \frac{\text{Total column load}}{\text{Permissible ground bearing pressure}}$$

$$\text{Minimum pad size required} = \frac{1323.8 \text{ kN}}{200 \text{ kN/m}^2} = 6.62 \text{ m}^2$$

Normally, pad foundations are square, except when practical constraints – for example, the presence of obstructions – mean that they have to be rectangular. So, if a square pad has to have a minimum plan area of 6.62 m², the minimum length of one of its sides is the square root of 6.62, i.e. 2.57 metres.

We shall round up this value to 2.7 metres, for the following reason. The self-weight of the base also acts on the ground below of course. But we cannot calculate the weight of the base until we know its size. By rounding up the base side length to 2.7 metres, we are increasing the base size to allow for the extra load due to the base itself. Let's assume that the depth of the foundation is 0.5 metres. We then need to check that the actual ground-bearing pressure is less than 200 kN/m²:

Total load in column = 1323.8 kN (calculated above)

Weight of 2.7 m square base = 24 kN/m³ × 0.5 m × 2.7 m × 2.7 m
= 87.5 kN

Total Load = 1323.8 + 87.5 = 1411.3 kN

So:

$$\text{Actual ground bearing pressure} = \frac{\text{Load}}{\text{Base area}} = \frac{1411.3 \text{ kN}}{2.7 \text{ m} \times 2.7 \text{ m}}$$

$$= 193.6 \text{ kN/m}^2$$

As this is less than 200 kN/m², a pad foundation of plan dimensions 2.7 m × 2.7 m is satisfactory.

Example 24.4: Loads in timber joist flooring

Due to its relatively low strength, timber flooring tends to be used in domestic construction where loading is light and spans are comparatively short. Timber flooring comprises timber beams (or joists, as they are usually known) at fairly close centres (typically 400, 450 or 600 mm).

For example, if a timber floor comprises 50 × 200 (width × depth in millimetres) timber joists, spaced 400 mm apart, supporting 10 mm thick timber boarding, the load on every metre length of joist (see Fig. 24.3 (a)) is as calculated below. Assume an imposed load of 1.5 kN/m² (normal in domestic construction) and that the unit weight of softwood is 5.9 kN/m³ (see Appendix 1).

Self-weight of joist per metre length = (0.05 m × 0.2 m × 1.0 m
× 5.9 kN/m³)
= 0.059 kN

Self-weight of boarding per metre of joist = (1.0 m × 0.4 m × 0.01 m
× 5.9 kN/m³)
= 0.024 kN

Live load per metre length of joist = (1.0 m × 0.4 m × 1.5 kN/m²)
= 0.6 kN

So

Total load per metre length of joist = (0.059 + 0.024 + 0.6) = 0.683 kN

Now let's suppose we wanted to calculate the load per square metre of flooring. A square metre of flooring with joists at 400 mm spacing will

(a) Load per metre of joist   (b) Load per square metre

Fig. 24.3 Loads on timber joist flooring.

contain 1/0.4 = 2.5 metres of timber joist (see Fig. 24.3 (b)). The total load per square metre of flooring is calculated as follows:

Self-weight of 2.5 length of joist = (0.05 m × 0.2 m × 2.5 m
× 5.9 kN/m³)
= 0.148 kN

Self-weight of one square metre of boarding = (1.0 m × 1.0 m
× 0.01 m × 5.9 kN/m³)
= 0.059 kN

Live load on one square metre of boarding = 1.5 kN

Therefore

Total load per square metre of timber flooring = (0.148 + 0.059 + 1.5)
= 1.71 kN

From inspection it can be seen that the dead load for timber flooring is usually a lot less than the associated live load.

The dead load part of the above calculation is (0.148 + 0.059) = 0.207 kN/m².

In general, a total dead load of 0.25 kN/m² is a convenient figure to use when performing calculations involving timber flooring.

Example 24.5: Loads due to steel beams

Structural steel beams come in standard sizes, each of which is designated by three figures multiplied together. The first figure is the nominal width, the second figure is the nominal overall depth and the third figure is the steel beam's own weight expressed in kg per metre length.

For example, the universal beam section designated 203 × 133 × 23 has a depth of approximately 203 mm, a width of around 133 mm and each metre of it weighs 23 kg.

Therefore, if we know the specific steel beam that is being used in a given situation, we know the self-weight in kg/m – which can be converted to kN/m by dividing by 100. For example, 23 kg/m = 0.23 kN/m.

If we don't know the specific steel beam size that is being used in a given situation, we can estimate the self-weight using the following rules of thumb:

- For steel beams up to 360 mm deep, self-weight (in kg/m) is about one-sixth of the depth (e.g. a 203 mm deep beam weighs 203/6 = 34 kg/m).
- For steel beams 360 to 800 mm deep, self-weight (in kg/m) is about one-quarter of the depth (e.g. a 533 mm deep beam weighs 533/4 = 133 kg/m).
- For steel beams over 800 mm deep, self-weight (in kg/m) is about one-half of the depth (e.g. a 914 mm deep beam weighs 914/2 = 457 kg/m).

Example 24.6: Loads on the supports of timber joist flooring

A timber floor comprises 50 × 200 timber joists spanning 4 metres from a central supporting beam to loadbearing walls on either side, as shown in Fig. 24.4. The central supporting beam, labelled 'Beam A', is a 203 mm deep steel beam and is supported by steel columns – labelled 'Column B' – at each end of its 6.0 metre span. If the dead load of the timber flooring is 0.25 kN/m² and the live load is 1.5 kN/m², calculate:

- the total load per metre length of beam A;
- the total load on each of the two columns B.

As the span of the flooring is 4.0 metres, half of that span (i.e. 2 metres) is supported by beam B. But it is 2 metres *on each side* of the beam, so the total portion of flooring supported by the beam = 2 × 2 m = 4 metres.

Total load per square metre of flooring = (0.25 + 1.5) = 1.75 kN/m²

Plan view of timber joist flooring

Fig. 24.4 Loads on supports of timber joist flooring.

Estimated self-weight of 203 mm deep steel beam (see example 24.5 above) = 203/6 = 34 kg/m
= 0.34 kN/m

Total load on beam B (per metre) from flooring = (4 m × 1 m
× 1.75 kN/m²)
= 7.0 kN

Self-weight of beam B per metre = 0.34 kN

Therefore

Total load per metre length of beam B = (7.0 + 0.34) = 7.34 kN

Total load on each supporting column = $\dfrac{7.34 \text{ kN} \times 6 \text{ m}}{2}$ = 22 kN

Figure 24.5 shows an atrium. Atriums are becoming increasingly common and feature large areas of glass (which may be horizontal, vertical or inclined). The glass needs to be supported and the supporting structure may be substantial.

Figure 24.6 shows the base of a modern high-rise office building. Note the inclined supports.

Fig. 24.5 Atrium, Learning Centre, Leeds Metropolitan University.

Fig. 24.6 Office building, Deansgate, Manchester.

Appendix 1

Weights of common building materials

A full list is given in British Standard BS 648. The more commonly used materials are discussed below. (Please note that these figures are 'typical' only, as the strength of any material varies according to the type or grade of material.)

Reinforced concrete

Unit weight: 24 kN/m³ (2400 kg/m³)
Therefore a 100 mm thick concrete wall weighs 2.4 kN/m² (240 kg/m²).

Blockwork

Unit weight: 22 kN/m³ (2200 kg/m³)
Therefore a 100 mm thick blockwork wall weighs 2.2 kN/m² (220 kg/m²).
Lightweight (aerated) blockwork can weigh considerably less (as low as 6 kN/m³).

Brickwork

Approximately the same weight as blockwork (see above).

Steel

Unit weight: 78.5 kN/m³ (7850 kg/m³)
Steel beams weigh between 0.2 and 2.0 kN/m (20–200 kg/m) depending on size.

Aluminium

Unit weight: 27.7 kN/m^3 (2771 kg/m^3)

Timber

Unit weight: Softwood: 5.9 kN/m^3 (590 kg/m^3). Hardwood: 12.5 kN/m^3 (1250 kg/m^3)

Therefore a 50mm x 200mm ('two by eight') softwood joist weighs 0.06 kN/m (6 kg/m).

Glass

Unit weight: 25 kN/m^3 (2500 kg/m^3)

Therefore the weight of glass is 0.025 kN (2.5 kg) per millimetre thick.

Water

Unit weight: 10 kN/m^3 (1000 kg/m^3)

Live loads

Live loads (i.e. non-permanent loads due to people and furniture in a room in a building) are assumed to be uniformly distributed and are expressed in kN/m^2. Values of live load depend on the use of the building (or part of the building) concerned. A full listing appears in British Standard BS 6399 Part 1. Some values are given below.

- Domestic: 1.5 kN/m^2
- Offices: 2.5 kN/m^2
- Cafes/restaurants: 2.0kN/m^2
- Classrooms: 3.0 kN/m^2
- Assembly: fixed seating: 4.0 kN/m^2
- Corridors/stairs in hotels, etc.: 4.0 kN/m^2
- Exhibitions: 4.0 kN/m^2
- Gyms: 5.0 kN/m^2
- Bars, concert halls, etc.: 5.0 kN/m^2
- Stages: 7.5 kN/m^2
- Shops: 4.0 kN/m^2
- Parking (cars): 2.5 kN/m^2
- Plant rooms: 7.5 kN/m^2.

*Note: British Standards can be viewed on the internet at *www.athens.ac.uk*. To access this site, an 'Athens password' is required, which can be obtained by students and staff at UK universities. See your university learning centre for details.

Appendix 2

Conversions and relationships between units

Inches, feet and metres

1 inch = 25.4 mm
1 foot = 304.8 mm = 0.3048 metres
1 metre = 3.281 feet
1 m^2 = 10.76 ft^2
1 ft^2 = 0.092 m^2

Yards and metres

1 yard = 3 feet = 36 inches = 0.9144 metres
1 metre = 1.094 yards
1 yd^2 = 0.836 m^2
1 m^2 = 1.196 yd^2

Acres and hectares

1 acre = 4840 yd^2 = 4047 m^2
1 hectare = 10,000 m^2 = 2.47 acres
1 acre = 0.405 hectares

Miles and kilometres

1 mile = 1760 yards = 1609.3 metres
1 km = 1000 metres
1 mile = 1.6093 km
1 km = 0.621 miles

Litres and cubic metres

1 metre = 100 cm
$1 \text{ m}^3 = 10^6 \text{ cm}^3$
1 millilitre = 1 cm^3
1 litre = 1000 millilitres = 1000 cm^3
1000 litres = 1 m^3

Pounds, kilograms and stones

1 lb = 0.454 kg = 454 g
1 kg = 2.203 lbs
1 stone = 14 lb = 6.356 kg

Kilograms, kN and tonnes

10 N = 1 kg
1000 N = 1 kN
10kN = 1 tonne = 1000 kg

Tons and tonnes

1 ton =160 stone = 1017 kg
1 tonne = 0.983 tons
1 ton = 1.017 tonnes

Appendix 3

Mathematics associated with right-angled triangles

Pythagoras' theorem

This states that 'the square of the hypotenuse of a right-angled triangle is equal to the sum of the squares of the other two sides'. In plain English this means that if the length of any two sides of a right-angled triangle are known, then the length of the third side can be determined using the relationships shown in Fig. A1.

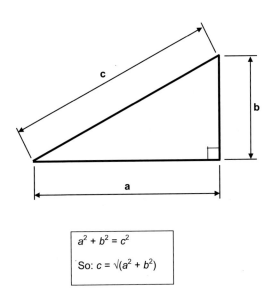

Fig. A1 Pythagoras' theorem.

Basic trigonometry

With respect to a right-angled triangle, sines, cosines and tangents (normally abbreviated to sin, cos and tan respectively) are defined in Fig. A2. There is nothing 'magic' about this. A sine, cosine or tangent is simply the ratio of the lengths of two sides of a right-angled triangle.

Suppose we are interested in the angle formed between two sides of the triangle. The angle is represented by the Greek letter θ and is measured in degrees. 'Hypotenuse' represents the length of the longest side of the right-angled triangle – which is always the side opposite the right angle. 'Opposite' represents the length of the side opposite the angle θ. 'Adjacent' represents the length of the side adjacent to the angle θ.

For example, if the 'opposite' side is 2 metres long and the 'hypotenuse' is 2.5 metres long, then sin θ = 2.0/2.5 = 0.8.

The reader should refer to a basic mathematics textbook if further information is required.

Fig. A2 Sines, cosines and tangents.

Appendix 4

Symbols

The units normally used in structural mechanics are given in brackets after each definition.

A = cross-sectional area (mm^2)
E = Young's Modulus or Modulus of Elasticity (kN/mm^2)
I = second moment of area (mm^4)
L = length; span of beam or slab (millimetres or metres)
M = moment (kN.m) (Chapter 8)
P = force (kN)
R = reaction (kN)
V = shear force (kN)
w = uniformly distributed load per metre (kN/m)
W = total uniformly distributed load per metre (kN)
σ = stress (direct or bending) (N/mm^2)
τ = shear stress (N/mm^2)

Index

aluminium 238
arches 22
area, second moment of 194
axial stress 181

beam 8
beam types 18
bending (definition) 16
bending, combined with axial stress 214–27
Bending, Engineer's Equation 194
bending moment 149
bending moment diagrams 152
bending stress 191

cable stayed structures 24
centroids 201
column 8, 19
components of force 56
compression 13
concrete 233–4, 255
contraflexure 163
cross bracing 95
cross section types 25

dead loads 36
density 32
determinacy, statical 83
direct stress 181
durability 232

eccentricity 218
elastic modulus 195
Engineers' Bending Equation 194
equilibrium 43–7

fixed supports 79
floors 7
force (definition) 10, 29
force, components of 56
forces, resultant of 53
force diagram (graphical) method 134–44
foundation, pad 21
foundation, raft 21
foundations, pressures under 225
foundation, strip 21
foundations 7, 21
framed structures 238

graphical method (for analysing pin jointed frames) 134–44

hogging 145
Hooke's Law 186

imposed loads 36
indeterminacy, statical 83

lateral loads 37
lateral stability 94

live loads 36
loads, dead 36
loads, lateral 37
loads, live 36
loads, calculation of 258–67
loads, nature of 38
load paths 40
long span structures 249

masonry 234–5, 255–6
mass 29
materials 229–41
mechanism 93
method of resolution at joints 105–125
method of sections 126–33
moments 11, 60–68

neutral axis 193
Newton (unit of force) 29

parallel axis theorem 204
perfect frame 92
pin jointed frames 100–104
pinned supports 78
pins 77
point load 39
portal frames 23

reaction (definition) 10
reactions 70–76
resolution of forces 51
restraints 80
resultants of forces 53
rigid joints 97
roller supports 78
roofs 6

safety 5
sagging 145
second moment of area 194
section modulus 195

serviceability 5
shear (definition) 17
shear force 148
shear force diagrams 152
shear stress 182
shear walls 94
slabs 18
span 249–57
span-to-depth ratio 256
stability 86–94
staircases 7
statical determinacy 83
statical indeterminacy 83
steel 236–7, 250–55
strain 183
stress (definition) 11
stress (introduction) 175–80
stress, axial 181
stress, bending 191
stress, combined bending and shear 214–27
stress, direct 181
stress, shear 182
support types 78
suspension structures 24

tension 14
theory of bending 191
timber 236, 255
trusses 23

uniformly distributed load (UDL) 40
uniformly varying load 40
unit weight 32
units of measurement 32

walls 6, 20
weight 30
wind loads 37

Young's modulus 186